Memoirs of the American Mathematical Society
Number 162

Richard L. Epstein

Minimal degrees of unsolvability and the full approximation construction

*"Perhaps I might tell more...
Outlines! I plead for my
brothers and sisters."*

Whitman, *Leaves of Grass*

Published by the
AMERICAN MATHEMATICAL SOCIETY
Providence, Rhode Island

VOLUME 3 · ISSUE 1 · NUMBER 162 · SEPTEMBER 1975 · CODEN: MAMCAU

ACKNOWLEDGEMENTS

This book was written as the author's Ph.D. dissertation at the University of California, Berkeley. Our advisors were Robert W. Robinson and Barry Cooper. We are grateful to both for their time and patience in teaching a recalcitrant student recursion theory. Their influence can be seen throughout this book.

A preliminary draft of Chapters I-IV was given as a series of informal lectures at the University of California, Berkeley in 1973. The comments of the participants were very helpful, especially those of S. Simpson.

The entire first draft was read by David Posner. His critical comments and observations have improved the text enormously. The motivations were especially clarified by his comments.

Many ideas have been incorporated from conversations we held with Len Sasso. We owe much, too, to his enthusiasm and his encouragement. Many thanks to a friend indeed.

Dale Ogar did the typing. Her help in laying out the manuscript is appreciated.

AMS (MOS) subject classification (1970). 02F30

Library of Congress Cataloging in Publication Data CIP

Epstein, Richard L 1947-
Minimal degrees of unsolvability and the full approximation construction.

(Memoirs of the American Mathematical Society ; no.162)
Originally presented as the author's thesis, University of California, Berkeley.
Bibliography: p.
1. Unsolvability (Mathematical logic) 2. Recursive functions. 3. Constructive mathematics. I. Title. II. Series: American Mathematical Society. Memoirs ; no. 162.
QA3.A57 no. 162 [QA248.54] 510'.8s [511'.3] 75-20308
ISBN 0-8218-1862-7

TABLE OF CONTENTS

ABSTRACT

A degree of unsolvability $\underline{m}$ is minimal if $\underline{m}$ is not recursive and there is no non-recursive $\underline{a} < \underline{m}$. In this Memoir we examine a method for constructing minimal degrees, namely the full approximation construction. Throughout we use a textbook approach with extensive motivations and diagrams. Emphasis is on techniques and their wide applicability.

Contents : Chapter I - notation and a standard construction of a minimal degree.

Chapter II - the full approximation construction of a minimal degree $\underline{m} < \underline{0}'$, the degree of the halting problem. It proceeds by constructing β_s, a finite string of 0's and 1's, recursive in s. Then $\underline{m}$ = the degree of B, where $B(x) = \lim_s \beta_s(x)$, $\forall x$. The construction is due to S.B. Cooper, though there is an earlier construction in this style due to C.E.M. Yates.

Chapter III - we show how to attach a technique called "followers", due to Cooper, to get a minimal degree $\underline{m}$ such that $\underline{m}' = \underline{0}'$ (due originally to Yates). This technique is useful in attaching any finite injury argument to the basic construction.

Chapter IV - the construction of Chapter III is expanded to get a theorem of Cooper that $\underline{c} \geq \underline{0}' \rightarrow$ there is a minimal degree $\underline{m}$ such that $\underline{m}' = \underline{c}$.

Chapter V - we give a new proof of a theorem of Yates that $\underline{0} < \underline{a}$ recursively enumerable $\rightarrow$ there is a minimal degree $\underline{m} < \underline{a}$ r.e. This establishes how to use permitting with respect to the basic construction.

Chapter VI - we sketch a method of R.W. Robinson that allows us, given $\underline{0} < \underline{a} < \underline{0}'$, to find $\underline{b} < \underline{0}'$ such that $\underline{b} \cup \underline{a} = \underline{0}'$. Here "$\cup$" is the join operation in the upper-semi-lattice of degrees. We then wed this to the basic construction to prove that $\underline{0} < \underline{a}$ r.e. $\rightarrow$ there is a minimal degree $\underline{m} < \underline{0}'$ such that $\underline{m} \cup \underline{a} = \underline{0}'$. As a corollary we have : $\underline{a}$ r.e.$\rightarrow$there is a complement for $\underline{a}$ in the degrees $\leq \underline{0}'$. We survey other work on complements in the degree $\leq \underline{0}'$.

INTRODUCTION

A minimal degree of unsolvability $\underline{m}$ is a degree such that $\underline{m} \neq \underline{0}$, the degree of recursive functions, but $\underline{a} < \underline{m} \rightarrow \underline{a} = \underline{0}$. Minimal degrees were first exhibited by C. Spector in 1956 [12]. Questions naturally arose as to what other conditions they could be made to satisfy. G.E. Sacks [8] showed that a minimal degree $\underline{m} < \underline{0}'$ could be constructed. Here $\underline{a}'$ denotes the degree of the halting problem of functions recursive in $\underline{a}$.

Sacks' construction proceeded by constructing at stage s+1 an initial segment of his set, $\beta_{s+1} \supset \beta_s$, where β_{s+1} was recursive in β_s and $0'$. But many questions concerning the relationships of degrees $\leqslant \underline{0}'$ and minimal degrees could still not be answered using this method. It became clear that a more flexible construction was needed.

In [1] S.B. Cooper devised a method for constructing a minimal degree $\underline{m} < \underline{0}'$ by a full approximation to it. That is, he constructed at stage s an initial segment of a characteristic function β_s; the degree $\underline{m}$ was then taken to be the degree of B, where $B(x) = \lim_s \beta_s(x)\ \forall x$. It is this full approximation to a minimal degree that we examine in this book.

Due to its flexibility this construction can be used in many contexts. In fact, the theorems which already have been proved using it have expanded considerably the study of the degrees $\leqslant \underline{0}'$. Many questions concerning the degrees $\leqslant \underline{0}'$ have flowed naturally from these theorems. Thus by examining the full approximation construction we examine the nature of the degrees $\leqslant \underline{0}'$ as well.

We present the full approximation in a variety of situations. We build consistently from relatively easy constructions to more complex ones. We do this by reproving many theorems already in the literature. This allows us to isolate various techniques of the study of the degrees $\leqslant \underline{0}'$, e.g. permitting, finite injury, etc., and show how they can be attached to the full approximation construction.

A second benefit of this approach is that certain very messy proofs which recur in many full approximation constructions can be done in complete detail in the simplest constructions. Later proofs can then be seen as modifications of these easier ones, and can be proved by abbreviated comments and references to the appropriate chapter. By thus isolating techniques, and the relevant difficulties of each, we hope to make the full approximation accessible to the recursion theorist as a working tool.

To this end we have tried to make the constructions as intuitive as possible. Each chapter begins with a long section entitled "Motivation" in which the nature of the problem and the ideas necessary to its solution are presented in detail. It is intended that these be read very carefully before the reader approaches the construction. We hope that in many cases the reader will understand the construction without having to plod through the proof. We include many diagrams too. They are a powerful teaching tool, one altogether too long neglected by recursion theorists.

In Chapter I we present the background and notation for the book by proving that there is a minimal degree. We suppose the reader has some idea of what degrees of unsolvability are. A passing acquaintance with priority arguments will prove very useful, but not essential. We try to explain each technique we use as we proceed so that no specific knowledge of the study of degrees $\leqslant \underline{0}'$ is required of the reader.

Chapter II ($\underline{m} < \underline{0}'$) presents a "bare bones" full approximation construction -- namely a minimal degree $\underline{m} < \underline{0}'$. This is the hardest chapter, but it is basic to all the others. The mastery of this one will make the other constructions appear natural.

We then prove a theorem of Cooper that $\underline{c} \geqslant \underline{0}' \to$ $\underline{m}$ minimal such that $\underline{m}' = \underline{c}$. This we do in two parts. The first, Chapter II ($\underline{m}' = \underline{0}'$) gives a construction of a minimal degree $\underline{m}$ such that $\underline{m}' = \underline{0}'$. This is a good illustration of how to attach a finite injury argument to the full approximation by way of "followers." In the second part, Chapter IV ($\underline{m}' = \underline{c} \geqslant \underline{0}'$) we show how to expand the construction of $\underline{m}' = \underline{0}'$ to a "tree of trees" to obtain the theorem. This introduces a method for using the full approximation in general degree theory, by constructing uncountably many sets at once. Several open problems are posed.

In Chapter V ($\underline{m} < \underline{a}$ r.e.) we prove a theorem of C.E.M. Yates that given $\underline{0} < \underline{a}$ recursively enumerable (r.e.) there is a minimal degree $\underline{m} < \underline{a}$. The proof differs significantly from Yates' [13] because we base it on the full approximation. This construction serves as an archetypal application of "permitting" to the full approximation. "Permitting" which we explain in Chapter V, is a way to construct degrees with respect to r.e. degrees.

Lastly in Chapter VI ($\underline{m} \cup \underline{a}$ r.e. $= \underline{0}'$) we prove a new theorem that given $\underline{0} < \underline{a}$ r.e. there is a minimal degree $\underline{m} < \underline{0}'$ such that $\underline{m} \cup \underline{a} = \underline{0}'$. Here "$\cup$" indicates the "join" in the upper semi-lattice of degrees. The method of proof is an amalgamation of the permitting techniques of Chapter V with a technique of R.W. Robinson [6] that given $\underline{0} < \underline{a} < \underline{0}'$ there is a degree $\underline{b} < \underline{0}'$ such that $\underline{b} \cup \underline{a} = \underline{0}$. We outline this "join technique" in some detail first. As a corollary we have that if $\underline{a}$ is r.e. then $\underline{a}$ has a complement in the degrees $\leqslant \underline{0}'$. We survey the other work on complements for degrees $\leqslant \underline{0}'$ and suggest further problems.

All proofs are indented and end with the symbol □. An index is also provided.

We have attempted throughout to give credit where credit is due for each result we mention. Any errors or omissions of references are not intentional.

Dick Epstein
Berkeley, California 1973

CHAPTER I: A MINIMAL DEGREE

Sections

* * *

The purpose of this Chapter is to establish notation and give a simple proof of the existence of a minimal degree. Minimal degrees were first shown to exist by Spector (12). Most of this Chapter comes from notes of a seminar given by S.B. Cooper in 1971 at the University of California, Berkeley.

Received by the editor May 20, 1974

Functionals and Trees

Let $\mathbb{N}$ denote the natural numbers.

A,B,C,D will always denote sets of natural numbers.

If $A \subseteq \mathbb{N}$, we have the characteristic function of A:

$$A(x) = \begin{cases} 0 & \text{if } x \notin A \\ 1 & \text{if } x \in A \end{cases}$$

There will be no difficulty in distinguishing between the set A and the function A(x).

Definition: A string is the restriction of a characteristic function A to a finite initial segment of A.

We write A[n] for A restricted to $x \leq n$

$\emptyset$ denotes the empty string.

Unless otherwise indicated, small Greek letters represent strings.

A string is just a finite ordered set of 0's and 1's.

Definition: The length of a string σ is:

$lh(\sigma) = n + 1$ if $\sigma = A[n]$ some A,n.

$lh(\emptyset) = 0$.

We may write $\sigma[n]$ for σ restricted to $x \leq n$ if $lh(\sigma) \geq n + 1$.

1,0 are used for the strings of length 1, with values 1, 0 respectively.

We write $\sigma \subset \tau$ (or $\sigma \subsetneq \tau$ for emphasis) if $\sigma = \tau[n]$ some n and $\sigma \neq \tau$.

We write $\sigma \subseteq \tau$ for $\sigma \subset \tau$ or $\sigma = \tau$.

We write $\sigma \wedge \tau$ for the longest γ such that $\gamma \subseteq \sigma$ and $\gamma \subseteq \tau$.

I-1

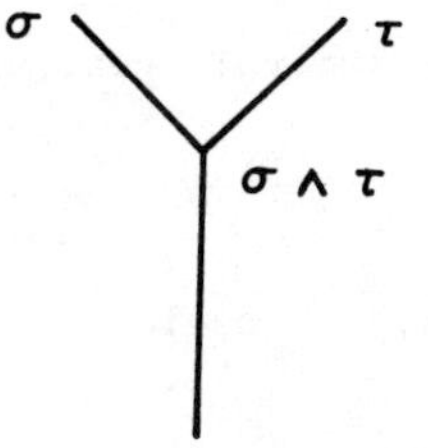

<u>Definition</u>: The <u>ordering of strings</u> we use is

$$\sigma \text{ is less than } \tau \text{ if } \begin{cases} lh(\sigma) < lh(\tau) \\ \quad\text{or} \\ lh(\sigma) = lh(\tau) \text{ and} \\ \quad \mu x(\sigma(x) \neq \tau(x)),\ \sigma(x) = 0. \end{cases}$$

$$(\sigma_0,\sigma_1) \text{ is less than } (\tau_0,\tau_1) \text{ if } \begin{cases} \sigma_0 \text{ is less than } \tau_0 \text{ and } \tau_1 \\ \quad\text{or} \\ \sigma_0 = \tau_i,\ \sigma_0 \text{ is less than } \tau_{1 \dot{-} i} \\ \quad\text{and } \sigma_1 \text{ is less than or} \\ \quad\text{equal to } \tau_{1 \dot{-} i}. \end{cases}$$

All pairs (σ_0,σ_1) such that σ_0 is less than or equal to σ_1 are comparable by this ordering.

<u>Definition</u>: Let $lh(\sigma) = s$, $lh(\tau) = t$ then σ <u>concatenated with</u> τ, <u>$\sigma * \tau$</u>, is:

$$(\sigma * \tau)(x) = \begin{cases} \sigma(x) & \text{if } x < s \\ \tau(x - s) & \text{if } s \leqslant x < t \end{cases}$$

e.g. $\sigma = (010)$, $\tau = (111)$

$\sigma * \tau = (010111)$

Definition: Φ is a partial recursive functional

if it is an effectively generable set of triples of the form:

$$\langle \sigma,x,y\rangle, \text{ called axioms}$$

such that

if (σ,x,y') and $(\sigma',x,y) \in \Phi$

and $\sigma \subset \sigma'$

then $y = y'$.

(The axioms are consistent)

$\Phi(A,x) = y \iff$ for some beginning $A[n]$ of A we have

$$\langle A[n],x,y\rangle \in \Phi.$$

This definition is acceptable since the axioms are consistent.

$\Phi(A,x)$ is a partial recursive function.

We will often write $\Phi(A)(x)$ for $\Phi(A,x)$.

Let $\{\Phi_e : e \geqslant 0\}$ be a standard enumeration of 0-1 valued partial recursive functionals, based on a standard enumeration of the recursively enumerable sets, say.

Let $\Phi_{e,s}$ be Φ_e enumerated to stage s:

$$\Phi_{e,s}(A,x) = y \iff x < s,\ y < s \text{ and some } n < s,\ \langle A[n],x,y\rangle \in \Phi.$$

We take $\phi_n = \Phi_n(\mathbb{N})$ as our enumeration of partial recursive functions, and $\phi_{n,s} = \Phi_{n,s}(\mathbb{N})$.

We write $\sigma \subseteq \phi_{n,s} \iff \forall x < lh(\sigma),\ \sigma(x) = \phi_{n,s}(x)$, and the latter is defined.

Definition: A is recursive in B, written $A \leqslant_T B \iff$ some e,

$\Phi_e(B) = A$.

That is $\forall x \;\; \Phi_e(B)(x) = A(x)$.

It is clear that this definition is equivalent to all the usual definitions of Turing reducibility.

The Turing degrees, or degrees of unsolvability, are the equivalence classes defined by $\equiv_T$.

We write

$$\deg(A) = \underline{A} = \underline{a}$$

for the degree of A, and

$\underline{0}$ for the degree of $\mathbb{N}$.

Notation: If ϕ is a function, defined on any domain, we write

$\phi(x)\downarrow$ for "$\phi(x)$ is defined (converges)."

$\phi(x)\not\downarrow$ for "$\phi(x)$ is not defined."

We write $\phi(x) \cong \theta(x)$ if both sides are defined and equal, or both undefined.

If we construct a function f by stages, at s we may have $f_s(x)$ not yet defined. This is different than if $f(x)$ diverges, which means at all stages, $f_s(x)\not\downarrow$ or $f_s(x) \neq f_{s+1}(x)$ infinitely often.

We sometimes read "$\phi(x)\downarrow$" as "$\phi(x)$ is defined by".

We write $0'$ for the set $\{x : \phi_x(x)\downarrow\}$.

If $0^{(n)}$ is defined we set

$$0^{(n+1)} = \{x : \Phi_x(0^{(n)}(x)\downarrow)\}$$

We usually write $0''$ for $0^{(2)}$.

Definition: T is a tree if it is a partial function from strings into strings such that

(i) If $T(\sigma)$, $T(\sigma')\downarrow$ and $\sigma \subset \sigma'$ then $T(\sigma) \subset T(\sigma')$

(ii) If $T(\sigma * i)\downarrow$ $i = 0$ or 1, then $T(\sigma * (1 \dot{-} i))\downarrow$ and they are not equal.

(iii) If $T(\sigma)\downarrow$ and $\sigma' \subset \sigma$, then $T(\sigma')\downarrow$.

We will often identify T with its range, and say σ lies on T to mean $\exists\, \tau$, $T(\tau) = \sigma$.

We write $T(\sigma * 0)(\sigma * 1)$ for $T(\sigma * 0)$, $T(\sigma * 1)$. The identity tree is T such that $T(\sigma) = \sigma, \forall \sigma$.

All our trees grow upward, nature's way. Thus $T(\emptyset)$ is the lowest string on T, and if $\sigma \subset \tau$ then

$T(\sigma)$ lies below $T(\tau)$,

$T(\tau)$ lies above $T(\sigma)$;

we use higher and lower similarly.

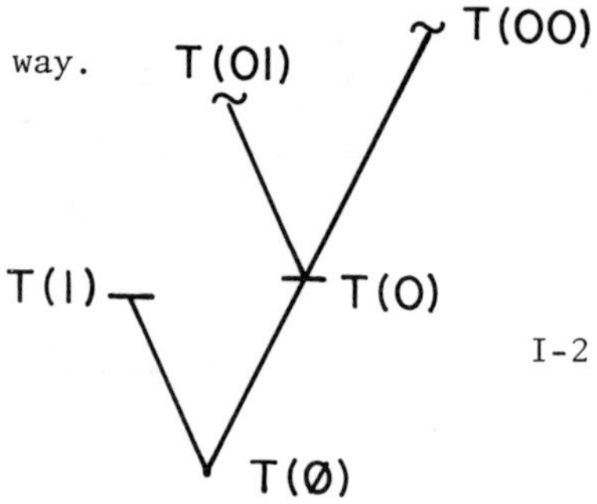

I-2

We say $T(\sigma)$ passes through x if $T(\sigma)(x)\downarrow$;

$T(\sigma)$ passes through τ if $T(\sigma) \supseteq \tau$.

I-3

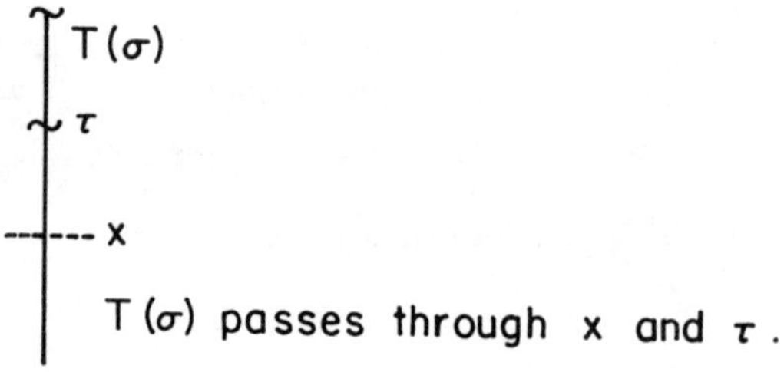

We say τ extends $T(\sigma)$ if $\tau \supsetneq T(\sigma)$,

and

T', a tree, has an extension lying above $T(\sigma)$ if some τ,

$T'(\tau) \supsetneq T(\sigma)$.

Definition: $\{\sigma_1, \sigma_2\}$ is a branching on T

if some τ,

$$\{T(\tau * 0)(\tau * 1)\} = \{\sigma_1, \sigma_2\}$$

We often refer to such $T(\tau)$ as a node.

Definition: T' is a subtree of T, written $T' \subseteq T$.
if T' is a tree and range $T' \subseteq$ range T

Definition: T is a partial recursive tree if T is a tree and is a partial recursive function.

Definition: T is full if $\forall \tau$, $T(\tau)\downarrow$.

Definition: The full subtree of T above π, $F_\pi(T)$ is:

$$F_\pi(T)(\emptyset) = T(\pi)$$

$$F_\pi(T)(\tau) = T(\pi * \tau).$$

If T is a partial recursive tree, its range is recursively enumerable. We write T_s for the largest full subtree of T above $\emptyset$ generated by stage s. Then $T = \bigcup_{s \geqslant 0} T_s$.

If T is partial recursive then every full subtree of T is also partial recursive.

Splitting Trees and Computation Lemmas

Definition: σ_1, σ_2 e-split σ through x

(split σ for e through x)

if

$\sigma_1 \wedge \sigma_2 \supset \sigma$ and

$\Phi_e(\sigma_1, x)\downarrow \neq \Phi_e(\sigma_2, x)\downarrow$

σ_1, σ_2 e-split σ through x at stage s

if the above holds with $\Phi_{e,s}$ for Φ_e. Note that this requires $x < s$.

Definition: T is an e-splitting tree

(splitting tree for e)

if

$\forall \tau,\ T(\tau * 0)(\tau * 1)\downarrow$

$\Rightarrow T(\tau * 0)(\tau * 1)$ e-split $T(\tau)$, through some x.

Definition: $Sp_{\pi,e}(T)$, the e-splitting subtree of T above π, is:

$Sp_{\pi,e,s}(T)(\emptyset) = T_s(\pi)$

$Sp_{\pi,e,s-1}(\tau)\downarrow$ then $Sp_{\pi,e,s}(\tau) = Sp_{\pi,e,s-1}(\tau)$.

$Sp_{\pi,e,s}(T)(\tau)\downarrow$ then

$Sp_{\pi,e,s}(T)(\tau * 0)(\tau * 1) = \mu(\sigma_1, \sigma_2)$ such that σ_1, σ_2 lie on T_s and σ_1, σ_2 e-split $Sp_{\pi,e,s}(\tau)$ at stage s;

$$\bigcup_{s \geq 0} Sp_{\pi,e,s}(T) = Sp_{\pi,e}(T)$$

If T is partial recursive, then so is $Sp_{\pi,e}(T)$. For any tree T, $Sp_{\pi,e}(T)$ is a subtree which is an e-splitting tree. If T is infinite (i.e. range T is infinite) $Sp_{\pi,e}(T)$ is not necessarily infinite.

Definition: A is on a tree T if the range of T contains infinitely many beginnings of A.

Note that if A lies on T and $T(\sigma) \subset A$ then $T(\sigma * i) \subset A$ $i = 0$ or 1.

Lemma 1: Let T be a partial recursive tree, A on T,

if $\nexists \sigma_1, \sigma_2$ lying on T such that σ_1, σ_2 e-split $T(\emptyset)$,

then

$\Phi_e(A)$ is recursive if it is total.

Proof: To compute $\Phi_e(A)(x)$, look for a σ such that σ lies on T and $\Phi_e(\sigma, x)\downarrow$.

Then $\Phi_e(\sigma, x) = \Phi_e(A, x)$.

If they were not equal, then, since $\Phi_e(A, x)\downarrow$, $\exists$ m such that $A[m]$ lies on T and $\Phi_e(A[m], x)\downarrow$. $A[m], \sigma$ would then e-split, a contradiction.

Since T is partial recursive the computation is recursive. □

Lemma 2: Let T be a partial recursive tree,

A on T

and assume T is an e-splitting tree.

Then

$$\Phi_e(A) \equiv_T A, \text{ if } \Phi_e(A) \text{ is total.}$$

Proof: $\Phi_e(A) \leqslant_T A$ always

To compute A from $\Phi_e(A)$ we give a recursive procedure to develop infinitely many beginnings of A.

$T(\emptyset)$ is a beginning of A.

Suppose $T(\tau)$ is a beginning of A.

One of $T(\tau * 0)(\tau * 1)$ is then a beginning of A.

Let $i \in \{0,1\}$ be such that

$$\Phi_e(T(\tau * i))(x) = \Phi_e(A)(x)$$

for some x through which $T(\tau * 0)(\tau * 1)$ e-split.

Then $T(\tau * i) \subset A$, as

$$\Phi_e(T(\tau * 1 - i)) \not\subset \Phi_e(A). \quad \square$$

A Minimal Degree

There is a minimal degree $\underline{m}$ (Spector)

We construct a set B of minimal degree by stages. At stage n+1 we will construct a tree T_{n+1} with T_{n+1} a subtree of T_n and $T_{n+1}(\emptyset) \supsetneq T_n(\emptyset)$. We will have $B = \bigcup_{n\geq 0} T_n(\emptyset)$.

Stage 0: T_0 = identity tree

Stage n+1: Let x be the least number such that

$$T_n(0)(x) \neq T_n(1)(x).$$

See if $\phi_n(x)\downarrow$.

If $\phi_n(x)\not\downarrow$ then $\sigma^{n+1} = T_n(0)$

If $\phi_n(x)\downarrow$ let i be minimal, $i \in \{0,1\}$ such that $T_n(i)(x) \neq \phi_n(x)$. Set $\sigma^{n+1} = T_n(i)$.

See if $\exists$ a string π lying on T such that $\pi \supset \sigma^{n+1}$ and such that no pair of strings lying on T above π n+1-splits.

Case I: If $\exists\pi$ set π_{n+1} = least such π.

Define

$$T_{n+1} = F_{\pi_{n+1}}(T_n).$$

Case II: If $\nexists\pi$, let $\pi_{n+1} = \sigma^{n+1}$ and define

$$T_{n+1} = Sp_{\pi_{n+1},}(T_n).$$

Let $B = \bigcup_{n\geq 0} T_n(\emptyset)$. That is $B(x) = T_n(\emptyset)(x)$ such that n is minimal, $T_n(\emptyset)(x)\downarrow$.

B lies on T_n $\forall n$ as $T_{n+1} \subset T_n$.

T_n is partial recursive $\forall n$, though not uniformly in n.

Lemma 1: B is not recursive.

Proof: B recursive => B = ϕ_{n+1} some n,

=> B $\not\supset$ $T_{n+1}(\emptyset)$, a contradiction. □

Lemma 2: $\forall n > 0$, $\Phi_n(B)$ total

=>

$\Phi_n(B)$ recursive or $\Phi_n(B) \equiv_T B$.

Proof: Given n, B lies on T_n

T_n satisfies the hypotheses of one of the computation lemmas. So $\Phi_n(B)$ must satisfy the conclusion of one of the computation lemmas. □

Hence b is minimal. □

Note: In this proof, as in all following proofs, we will ignore ϕ_0, Φ_0, so that T_{n+1} is constructed with reference to n+1-splittings. Since every recursive function has infinitely many indices we lose no generality.

Corollary: There is a minimal degree $\underline{m} < \underline{0}''$

Proof: By inspection, in the above construction we asked only questions recursive in 0''. So $B \leqslant_T 0''$. $B \neq_T 0''$ since 0'' is not minimal. □

We will refer to this construction as $\underline{0} < \underline{m}$.
$\underline{m}$ will always denote a minimal degree.

CHAPTER II: A MINIMAL DEGREE $\underline{m} < \underline{0}'$

Sections

*   *   *

It was first shown by Sacks that there is a minimal degree $\underline{m} < \underline{0}'$. His construction was introduced in (8).

That method takes the Spector argument and, instead of asking questions recursive in $0''$, constructs β_{s+1}, a finite string, at stage $s + 1$ asking only questions recursive in β_s and $0'$. Then with $\beta_{s+1} \supsetneq \beta_s$, $B = \bigcup_{s>0} \beta_s \leqslant_T 0'$. The construction approximates the splitting trees all at once. No particular attention is given to $T_e(\tau)$ for any specific τ, other than $\emptyset$. We refer the reader to Yates (13) for a clear presentation of that construction, by what is known as the e-state method.

Cooper, trying to solve the problem of whether there are a pair of minimal degrees $\underline{m}_1, \underline{m}_2 < 0'$ such that $\underline{m}_1 \cup \underline{m}_2 = \underline{0}'$, found that the e-state method was not sufficiently flexible. In the process of solving that problem in the affirmative (1), he devised a new construction of a minimal degree $\underline{m} < \underline{0}'$. That construction is called the full approximation to a minimal degree, because in it only recursive questions are asked. $\{\beta_s\}_{s>0}$ is uniformly recursive, though not monotone. $B = \lim_s \beta_s \leqslant_T 0'$ then. That the limit of uniformly recursive functions is recursive in $0'$ follows from the Limit Lemma of Shoenfield [11 , page 29.] In the full approximation, $T_e(\tau)$, for every e, τ, is separately approximated. This yields much greater flexibility and application, though in the case of simply exhibiting a minimal $\underline{m} < \underline{0}'$ the proof is much longer than the e-state method.

In this chapter we construct a minimal degree $\underline{m} < \underline{0}'$ by the full approximation method. We have attached no other conditions to that minimal degree since it is imperative that the basic construction be clear to the reader. All other constructions below $\underline{0}'$ in this work are expansions and modifications of this one; the mastery of this chapter is essential to further work on minimal degrees below $\underline{0}'$.

The reader should not be dismayed by the length of this chapter. Many details have been included which normally would be left to him to prove; by doing this we hope that his first encounter with the full approximation construction will be a pleasant, comprehensible one.

Motivation

We will set out to approximate the T_e and B of $\underline{0} < \underline{m}$. However, when we do this we shall find that the T_e we have constructed will not be partial recursive. At best we will get $T_e \leqslant_T 0'$. Once the T_e are constructed, and we have our set B in hand, we will see how we may prune the T_e to trees T_e^* which are partial recursive and play the role of the T_e in $\underline{0} < \underline{m}$.

So we wish to approximate the T_e of $\underline{0} < \underline{m}$. To begin with we'll need a T_0. Since we're proceeding by stages we'll use at stage s

$$T_{0,s} = T_0, \text{ the identity tree, to level } s,$$

where,

Definition: T, a tree, to level s is

$$\{T(\tau)\colon \operatorname{lh}(T(\tau)) \leqslant s, \text{ or } \operatorname{lh}(T(\tau)) > s \text{ but } \operatorname{lh}(T(\tau')) < s,\ \tau' * i = \tau,\ i = 0 \text{ or } 1\}$$

Our choices for branches on $T_{e,s}$ $\forall e$ will come from $T_{0,s}$.

We will construct $T_{1,s}$ by looking for 1-splittings on $T_{0,s}$. To begin $T_{1,s}$, just in case we can't find any 1-splittings, we'll define $T_{1,s}(\emptyset) = T_{0,s}(0)$ and $T_{1,s}(0)(1)$ = least pair of strings on $T_{0,s}$ above $T_{1,s}(\emptyset)$ (i.e. $T_{0,s}(00)(01)$). We need $T_{1,s}(0)(1)\downarrow$ so that we may later diagonalize against $\phi_{1,s}$ on $T_{1,s}$. If we have $T_{1,s}(\tau)\downarrow$ and find a 1-splitting σ_1, σ_2 of $T_{1,s}(\tau)$ then we'll define $T_{1,s}(\tau * 0)(\tau * 1) = \sigma_1, \sigma_2$ and never change that approximation. If we do not find a 1-splitting we never define $T_{1,s}(\tau * 0)(\tau * 1)$. So $T_{1,s} = Sp_{1,T_{1,s}(\emptyset)}(T_{0,s})$.

In this way we'll get $T_1 = \bigcup_s T_{1,s}$ which is a 1-splitting tree. Leave for the moment how we choose B, but suppose we have it. What if T_1 is not infinite along the path of B? That is, B does not lie on T_1. We'll then have $B \supset$ an end string on T_1, say $T_1(\tau)$, where

<u>Definition</u>: $T(\tau)$ is an <u>end string</u> on T if $T(\tau)\downarrow$ and $T(\tau * 0)(\tau * 1)\not\downarrow$.

Since we do not define $T_1(\tau * 0)(\tau * 1)$ no 1-splittings occur above $T_1(\tau) \subset B$ on T_0. In that case we don't want T_1, but rather $T_1^* = F_{T_1(\tau)}(T_0)$. If B does lie on T_1, we'll have $T_1^* = T_1$. In either case we have a tree appropriate to the application of one of the computation lemmas.

To approximate T_2 we can't wait until we are done with T_1; that would involve the use of 0'. We must simultaneously construct $T_{2,s}$ with $T_{1,s}$. Our first reaction is to have $T_{2,s} \subset T_{1,s}$ and $T_2 \subset T_1$ as in <u>0</u> < <u>m</u>. But what if we have T_1 finite along the path of B? We could have 2-splittings on T_1^* above $T_1(\tau)$, an end string on T_1. e.g.

<u>II-1</u>

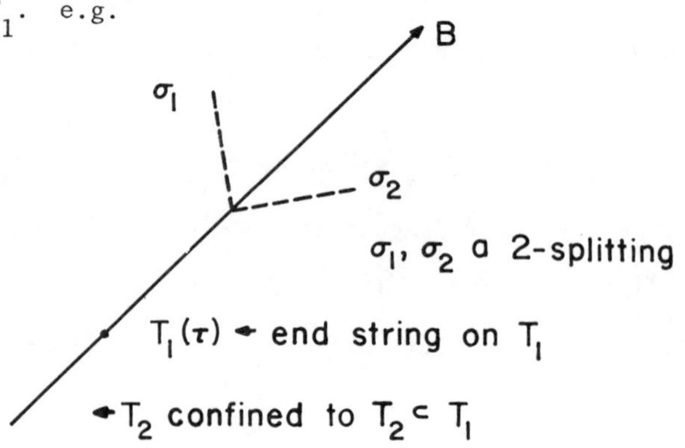

What we really want is $T_2 \subset T_1^*$. To get that we'll put branches on $T_{2,s}$ that either lie on $T_{1,s}$ or extend end strings on $T_{1,s}$.

Definition: σ is compatible with T if σ lies on T or σ extends an end string on T.

T' is compatible with T if every $T'(\tau)$ is compatible with T.

σ is compatible with τ if $\sigma \subseteq \tau$ or $\tau \subseteq \sigma$. We write $\sigma|\tau$ for "σ is incompatible with τ."

e.g.

II-2

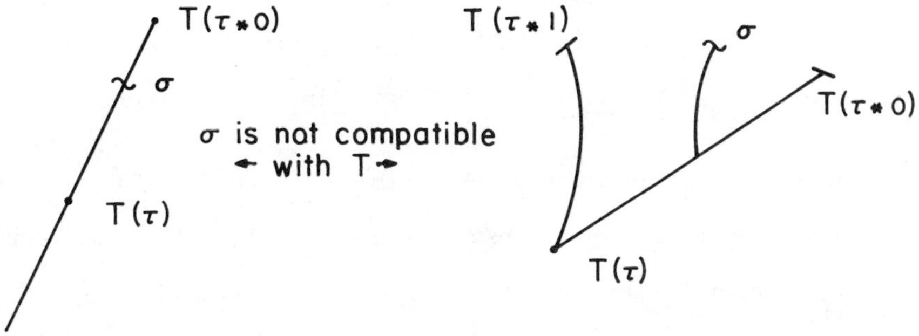

Note that T' compatible with T is not symmetric or transitive.

Now we begin $T_{2,s}(\emptyset) = T_{1,s}(0)$. To start off T_2, in case we can't find any 2-splittings, $T_{2,s}(0)(1)$ = least pair of branches lying on $T_{0,s}$ above $T_{2,s}(\emptyset)$ compatible with $T_{1,s}$.

Since the above phrase is used so often, we make a

Definition: (σ_1,σ_2) is a dummy extension of $T_{e,s}(\tau)$ if

$\sigma_1|\sigma_2$, they lie above $T_{e,s}(\tau)$ on $T_{0,s}$, and are compatible with $T_{e',s}$ $\forall e' < e$.

Continuing as with $T_{1,s}$ if $T_{2,s}(\tau)\downarrow$ and we can find a pair σ_1,σ_2 2-splitting $T_{2,s}(\tau)$ which are compatible with $T_{1,s}$ we'll define $T_{2,s}(\tau * 0)(\tau * 1) = \sigma_1,\sigma_2$. Otherwise we'll not define $T_{2,s}(\tau * 0)(\tau * 1)$.

We'll then define $T_2 = \lim_s T_{2,s}$, where by this we mean:

<u>Definition</u>: $\lim_s T_s(\tau)$ exists if

$$\exists s_0 \forall s > s_0 \; T_s(\tau) = T_{s_0}(\tau)\downarrow \text{ in which case}$$
$$\lim_s T_s(\tau) = T_{s_0}(\tau)$$

or

$$\exists s_1 \forall s > s_1 \; T_s(\tau)\uparrow, \text{ in which case } \lim_s T_s(\tau)\uparrow.$$

$$T = \lim_s T_s \text{ if } \forall\tau \; \lim_s T_s(\tau) \text{ exists} = T(\tau).$$

We'll beg the question of why $\lim_s T_{2,s}$ exists for now. However, one thing is clear: at best we can claim $T_2 \leqslant_T 0'$. We do not have T_2 partial recursive. As $T_{1,s}$ grows we will have to redefine branchings on $T_{2,s}$ if they extended what was, but is no longer, an end string on $T_{1,s}$.

How do we get a T_2^* partial recursive? Since what we want for T_2^* is T_2 relative to T_1^* we'll just ask what T_1^* is and then choose T_2^*. We define T_2^*: If T_2 is finite along the path of B, then set T_2^* = full subtree of T_1^* above some end string on T_2. If B lies on T_2 we'll begin $T_2^*(\emptyset)$ = first branch on $T_2 \subset B$ which lies above $T_1^*(\emptyset)$. So $T_2^*(\emptyset)$ will lie on T_1^*. Then we put branchings (hence 2-splittings) on T_2^* which appear on $T^*_{1,s}$ and $T_{2,s}$. We know that these branchings will be the permanent ones on T_2 because they either extend a permanent end string on T_1, or actually lie on T_1, since they lie on $T^*_{1,s}$. In this way we get a partial recursive $T_2{}^*$ appropriate to the application of one of the computation lemmas (either a 2-splitting tree, or no 2-splittings.)

We may continue in this manner defining $T_1, T_2, T_3, \ldots$. $T_{e,s}$ will be compatible with $T_{e',s}$ $\forall e' < e$. The $T_{e,s}$ will be uniformly recursive in e and s, and $T_e = \lim_s T_{e,s}$ will be recursive in $0'$. After we have all these trees, and B in hand, we'll prune them to $T_1{}^*, T_2{}^*, T_3{}^*, \ldots$. To define $T_e{}^*$ we will need to know how we define $T_{e'}^*$ $\forall e' < e$. This we may ask since each is partial recursive. $T_e{}^*$ is partial recursive, then, and appropriate to the application of one of the computation lemmas for e. Of course $\{T_e{}^*\}_{e>0}$ is not uniformly partial recursive.

The path of B will be, as in $\underline{0} < \underline{m}$, $B = \bigcup_{e>0} T_e(\emptyset)$. Since expressed that way we only have $B \leqslant_T 0''$, we actually choose B as $B(x) = \lim_s \beta_s(x)$, where $\beta_s = T_{m_s,s}(\emptyset)$, m_s maximal such that $T_{m_s,s}(\emptyset)\downarrow$. Since we'll have $T_{e+1}(\emptyset) \supsetneq T_e(\emptyset)$ the two definitions of B are equivalent. But in the latter we can see $B \leqslant_T 0'$. Of course to get B not recursive we'll diagonalize as we go along, as we did in $\underline{0} < \underline{m}$, so that $T_{e+1}(\emptyset) = T_e(i)$, $i \in \{0,1\}$ such that $T_e(i) \not\subseteq \phi_{e+1}$.

Note that it would not hurt our construction in any way to not erect e-splittings on $T_{e,s}$ above a string $T_{e,s}(\tau)$ such that $T_{e,s}(\tau) \not\subset \beta_s$ all sufficiently large s. (Diagram II-3). Splittings away from the final path of B will be superfluous. We'll use this fact not in this proof, but in most of our later ones to restrict ourselves at stage s to trying to e-split $T_{e,s}(\tau)$ only if

$T_{e,s}(\tau) \subset \beta_{s-1}$.

II-3

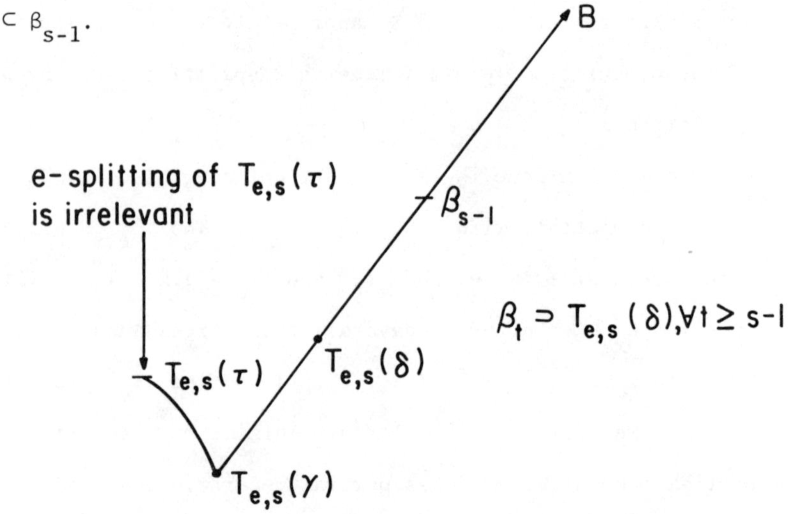

We have saved for the last the explanation of a basic device which we use to assure us that $\lim_s T_{e,s}(\tau)$ exists $\forall e,\tau$.

Boundary Strings and Dummy Extensions

Suppose we are at stage s and we have

II-4

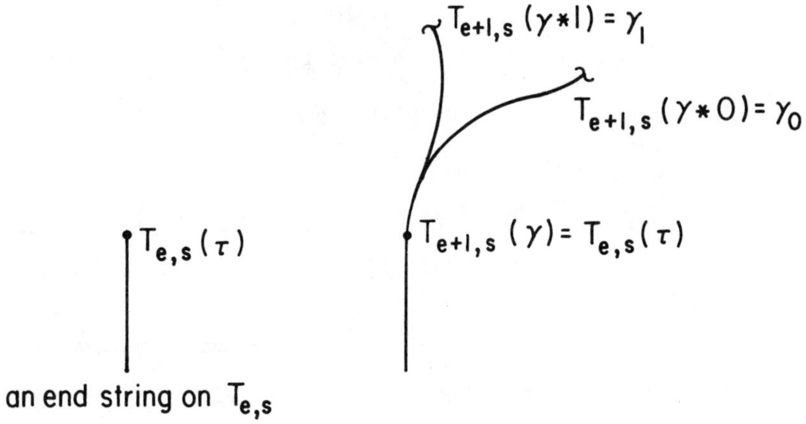

and at stage $s + 1$ we define $T_{e,s+1}(\tau * 0)(\tau * 1)$, an e-splitting pair,

II-5

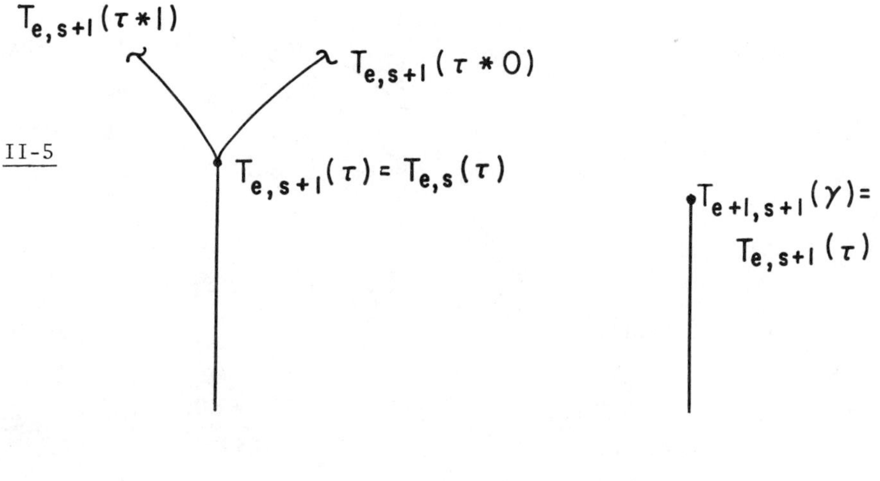

What will we choose as $T_{e+1,s+1}(\gamma * 0)(\gamma * 1)$? First we make a

<u>Definition</u>: If $T_{e,s}(\tau)\downarrow$ originally as part of an e-splitting pair but

$T_{e,s}(\tau * 0)(\tau * 1)\not\downarrow$, or are defined but are not an e-splitting pair,

then

$T_{e,s}(\tau)$ is a <u>boundary (bdry.) string for e</u> at s.

We will usually delete "at s" the stage being clear.

In II-5 $T_{e,s+1}(\tau * 0)(\tau * 1)$ are both boundary strings. They are end strings too. Shortly we shall see how a boundary string may not be an end string.

What to choose as $T_{e+1,s}(\gamma * 0)(\gamma * 1)$? Suppose

<u>II-6</u>

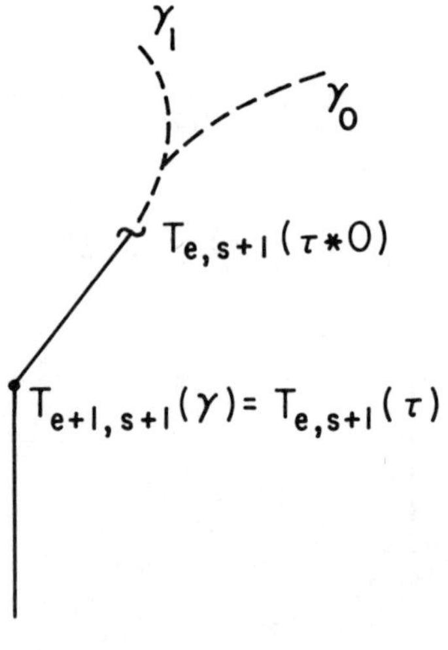

γ_0, γ_1 are both still compatible with $T_{e',s+1}$ $\forall e' < e+1$. The <u>obvious reaction</u> is to define $T_{e+1,s+1}(\gamma * 0)(\gamma * 1) = \gamma_0, \gamma_1$ again.

Or suppose

II-7

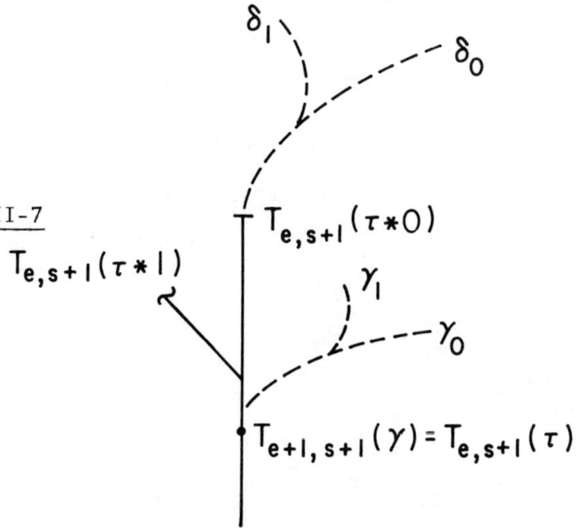

γ_0, γ_1 are no longer compatible with $T_{e,s+1}$ but there is now an e+1-splitting δ_0, δ_1 lying above $T_{e,s+1}(\tau * 0)$ compatible with $T_{e',s}$ $\forall e' < e + 1$. Here the obvious would be to define $T_{e+1,s+1}(\gamma * 0)(\gamma * 1) = \delta_0, \delta_1$.

Neither of these work.

Suppose we had the latter case and did set $T_{e+1,s+1}(\gamma * 0)(\gamma * 1) = \delta_0, \delta_1$. Nothing would prevent δ_0 or δ_1 from becoming incompatible with $T_{e',s+n}$ some $e' < e + 1$, $n > 1$, just as γ_0, γ_1 became incompatible. We might then redefine $T_{e+1,s+n}(\gamma * 0)(\gamma * 1) = \rho_0, \rho_1$. Again, ρ_0 or ρ_1 could become incompatible. We could keep chasing e +1-splittings above end strings on $T_{e',s+t}$, $e' \leqslant e$, indefinitely. That would prevent $T_{e+1,s}(\gamma * 0)(\gamma * 1)$ from going to a limit.

Let's use settle down to mean reach limits. That is, $T_{e,s}(\tau)$ has settled down if $\forall t > s$ $T_{e,t}(\tau) = T_{e,s}(\tau)$. $T_{e,s}$ has settled down to level r means $T_{e,s}(\tau)$ has settled down, $\forall \tau$ such that $T_{e,s}(\tau)$ is on T_e to level r.

How do we make $T_{e+1,s}(\gamma * 0)(\gamma * 1)$ settle down eventually? A boundary string for some $e' < e + 1$ (here $T_{e,s+1}(\tau * 0)$) intervenes between $T_{e+1,s+1}(\gamma)$ and an e+1-splitting $(\gamma_0, \gamma_1$ in II-6; δ_0, δ_1 in II-7). In this case we must choose

$$\begin{aligned} T_{e+1,s+1}(\gamma * 0)(\gamma * 1) &= T_{e,s+1}(\tau * 0)(\tau * 1) \\ &= \text{least dummy extension} \\ &\quad \text{of } T_{e+1,s+1}(\gamma) \end{aligned}$$

e.g. working from II-7

II-8

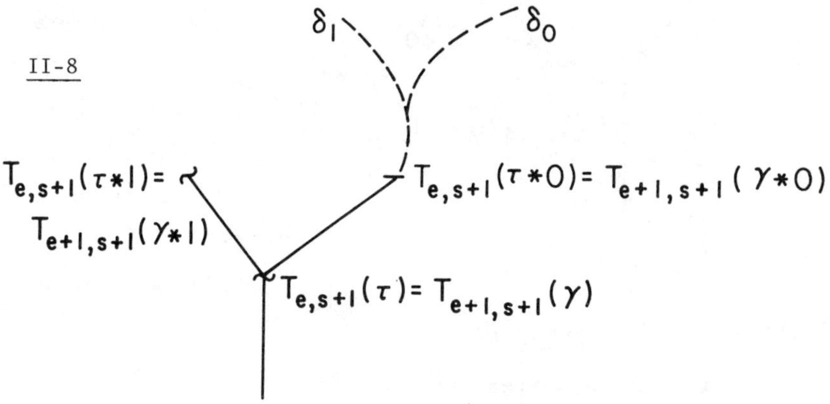

We say $T_{e+1,s}$ has captured the boundary strings $T_{e,s+1}(\tau * 0)(\tau * 1)$. We use the anthropomorphism "capture" to mean "erect as a branch," whether a dummy extension or not. $T_{e+1,s+1}(\gamma)$ is now a boundary string for $e + 1$. It is not an end string.

What does capturing boundary strings with dummy extensions do for us? Suppose $T_{e,s+1}(\tau)(\tau * 0)(\tau * 1)$ and $T_{e+1,s+1}(\gamma)$ have settled down.

If $T_{e,s+1}(\tau * 0)$ is an end string on T_e, then we have caused $T_{e+1,s+1}(\gamma * 0)(\gamma * 1)$ to settle down. But more, we have lifted $T_{e+1,s+1}$ above that end string on T_e. That is we may begin e+1-splitting above $T_{e,s+1}(\tau * 0) = T_{e+1,s+1}(\gamma * 0)$. This is essential, for if $B \supset T_e(\tau * 0)$ we must be allowed to e+1-split beginnings of B above $T_e(\tau * 0)$ as T_e^* will begin above $T_e(\tau * 0)$.

<u>If $T_{e,s+1}(\tau * 0)$ is not a permanent end string on T_e,</u> then we have two possibilities: We never find an e+1-splitting σ_0, σ_1 of $T_{e+1}(\gamma)$ compatible with $T_{e't}$ $\forall e' < e$, $t > s$, with no boundary string for any $e' < e + 1$ intervening. Then we'll never change $T_{e+1,s+1}(\gamma * 0)(\gamma * 1)$, as we never have a splitting to use instead. They have settled down. However, if we do find such a σ_0, σ_1 (diagram II-9) we'll use them as $T_{e+1,t}(\gamma * 0)(\gamma * 1) = \sigma_0, \sigma_1$. Since $T_{e+1,t}(\gamma)$ has settled down, and no boundary strings intervene, we'll have $T_{e+1,t}(\gamma * 0)(\gamma * 1)$ is permanent; that is they have settled down. This will become clear in the proof later.

<u>II-9</u>

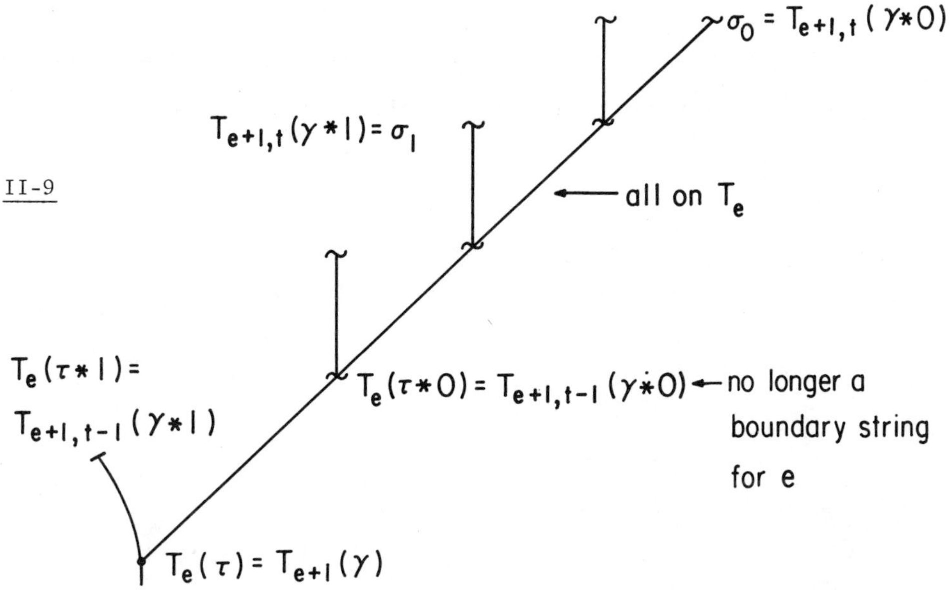

To summarize:

The purpose of a dummy extension is to lift $T_{e+1,s+1}$, say, above what appears to be an end string on $T_{e'}$, $e' < e + 1$; or equivalently it lifts $T_{e+1,s+1}$ above what appears to be the beginning of T_e^*. If it is a true end string on T_e, we will keep that dummy extension. If it is not, but we never find a suitable $e + 1$-splitting, we will keep that dummy extension anyway.

Dummy extensions facilitate the limiting process by assuring that if $T_{e,s}$ $(\tau * 0)(\tau * 1)$ are an e-splitting and are cancelled at stage $s+1$, then $T_{e,s}(\tau)$ must also be cancelled.

* * *

The reader should take note that our construction employs infinite injury. For each e, T_e may injure T_r, $r > e$, infinitely often. Our construction comes under control because each <u>branch</u>, $T_r(\tau)$, $r > e$, can be injured at most finitely often by T_e.

* * *

The motivation is done. We send the reader on his or her merry way down the path, hopefully minimal, of understanding the construction.

Construction of $\underline{m} < \underline{0}'$

Full approximation to $\underline{m} < \underline{0}'$

Stage 0: $T_{0,0} = \emptyset$, $\beta_0 = \emptyset$

Stage s: $T_{0,s}$ = identity tree to level s
$s > 0$

All strings mentioned at stage s lie on $T_{0,s}$

Suppose we have finished with $T_{e-1,s}$, $e > 0$.

If $T_{e-1,s}(0)$ $(1)\not\downarrow$ go to stage $s + 1$.

Otherwise we define $T_{e,s}$.

$T_{e,s}$:

$$T_{e,s}(\emptyset) = \begin{cases} T_{e-1,s}(0) & \text{if } \not\subseteq \phi_{e,s} \\ T_{e-1,s}(1) & \text{otherwise} \end{cases}$$

If $T_{e,s}(\tau)\downarrow = T_{e,s-1}(\tau)$ we proceed by cases to define $T_{e,s}(\tau * 0)(\tau * 1)$. If no case applies we do not define them.

If $T_{e,s}(\tau)\not\downarrow$, or $T_{e,s}(\tau)\downarrow \neq T_{e,s-1}(\tau)$ we do not define $T_{e,s}(\tau * 0)(\tau * 1)$.

Case I: $T_{e,s-1}(\tau * 0)(\tau * 1)\downarrow$ and compatible with $T_{r,s}$ $\forall r < e$

and

(i) $\tau = \emptyset$

or (ii) $T_{e,s-1}(\tau * 0)(\tau * 1)\downarrow$ originally by Case II and (b) of Case II applies still (mutatis mutandis).

or (iii) $\nexists \sigma_1, \sigma_2$ as in Case II.

Then set $T_{e,s}(\tau * 0)(\tau * 1) = T_{e,s-1}(\tau * 0)(\tau * 1)$.

Case II: Case I does not apply and $\tau \neq \emptyset$, and $\exists \sigma_1, \sigma_2$ such that

(a) σ_1, σ_2 are compatible with $T_{r,s}$ $\forall r < e$,

(b) every $\sigma' \subset \sigma_1$ or $\sigma' \subset \sigma_2$ which is a boundary string for some r, $r < e$, satisfies

$$\sigma' \subsetneq T_{e,s}(\tau),$$

and (c) σ_1, σ_2 e-split $T_{e,s}(\tau)$ at s.

Then set $T_{e,s}(\tau * 0)(\tau * 1) =$ least such σ_1, σ_2.

Case III: Cases I and II do not hold and either

(i) $\tau = \emptyset$

or (ii) $\exists \sigma_1, \sigma_2$ as in Case II except that (b) fails.

Let τ_0, τ_1 be the least pair of strings extending $T_{e,s}(\tau)$ compatible with $T_{r,s}$ $\forall r < e$, (the least dummy extension of $T_{e,s}(\tau)$).

Then set $T_{e,s}(\tau * 0)(\tau * 1) = \tau_0, \tau_1$.

When no more branches may be defined at stage s, we set

$$m_s = \max m(T_{m,s}(\emptyset)\downarrow),$$

and

$$\beta_s = T_{m_s,s}(\emptyset).$$

Proceed to stage $s + 1$.

This completes the construction.

*   *   *

Proof

Lemma 1: $\lim_s T_{e,s}(\tau)$ exists $\forall e, \tau$.

Proof: Recall that this means given e, τ,

$$\exists s_0, \forall s > s_0\ T_{e,s}(\tau) = T_{e,s_0}(\tau)\downarrow$$
$$\text{or} \quad \exists s_1, \forall s > s_1, T_{e,s}(\tau)\not\downarrow.$$

The proof is by induction, first on e, then on the $lh(\tau)$. It is immediate for T_0.

Suppose $\forall\tau \lim_s T_{e,s}(\tau)$ exists. We call $T_e = \lim T_{e,s}$.

$\lim_s T_{e+1,s}(\emptyset)\downarrow$:

It is clear that $\lim_s T_{e,s}(\emptyset) = \begin{cases} T_e(0) & \text{if } \not\subseteq \phi_e \\ T_e(1) & \text{otherwise.} \end{cases}$

$\lim_s T_{e+1,s}(0)(1)\downarrow$:

Let s_0 be such that $\forall s > s_0\ T_{e+1,s}(\emptyset) = T_{e+1,s_0}(\emptyset)$.

Let $s_1 \geqslant s_0$ be such that there is a unique pair σ_1, σ_2 such that $\forall s \geqslant s, \sigma_1, \sigma_2$ is the least dummy extension of $T_{e+1,s}(\emptyset)$. Such a stage exists because by the induction hypothesis $\lim_s T_{e',s}$ exists $\forall e' \leqslant e$. Indeed, we may define this stage:

let s_1 be such that $s_1 \geqslant s_0$ and $\forall e' \leqslant e$,

$\forall s \geqslant s_1 \quad T_{e',s}(\rho * i) \cong T_{e',s_1}(\rho * i)$, $i = 0$ or 1,

$\forall \rho$ such that $T_{e'}(\rho)\downarrow \subseteq T_{e+1}(\emptyset)$.

(We'll later use this argument by simply saying: "Let s be such that $\forall e' \leqslant e$, T_{e',s_1} has settled down to at least one branching above $T_{e+1}(\emptyset)$.")

Remark: We now point out an important fact:

For any τ, if δ_0, δ_1 is the least dummy extension of $T_{e,s}(\tau)$, then $\delta_0, \delta_1 = T_{e',s}(\delta * 0)(\delta * 1)$ $e' < e$, where:

e' is maximal and δ minimal thereafter

such that $T_{e',s}(\delta * 0) \wedge T_{e',s}(\delta * 1) \supseteq T_{e,s}(\tau)$.

Why? $T_{e',s}(\delta * 0)(\delta * 1)$ are compatible with $T_{r,s}$ $\forall r < e$. If $\{T_{e',s}(\delta * 0)(\delta * 1)\} \neq \{\delta_0, \delta_1\}$ then we must have:

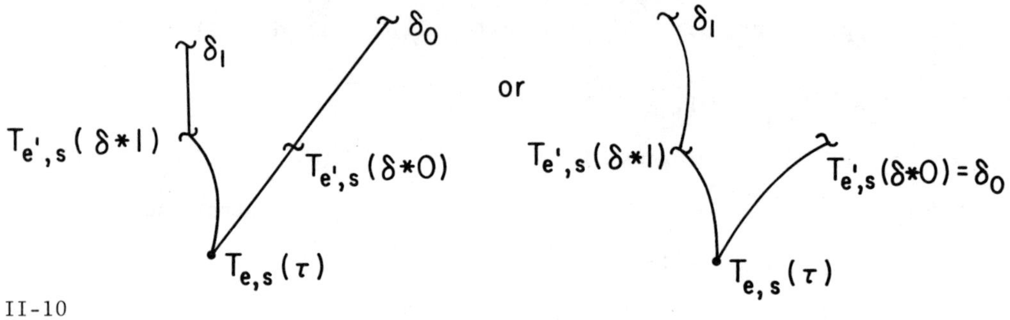

II-10

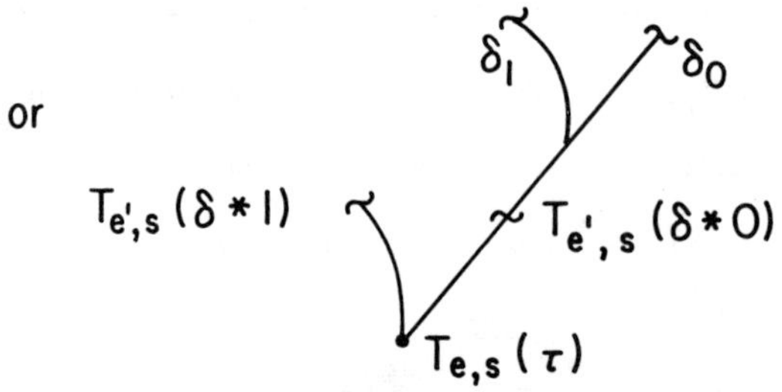

But then by our ordering (δ_0, δ_1) is not minimal.

If for some $t < s_1$, $\forall s > t, T_{e+1,s}(0)(1) = T_{e+1,t}(0)(1)$ we are done. If not, we must redefine $T_{e+1,s}(0)(1)$ at some stage $s \geqslant s_1$. But if $s \geqslant s_1$, we must define $T_{e+1,s}(0)(1) = \sigma_1, \sigma_2$, since we always define them by Case I or III. So $\lim_s T_{e+1,s}(0)(1)\downarrow = \sigma_1, \sigma_2$.

Suppose $\lim_s T_{e+1,s}(\tau)\downarrow$, $\tau \supset \emptyset$. We show that $\lim_s T_{e+1,s}(\tau * 0)(\tau * 1)$ exist.

Let s_0 be such that $\forall s > s_0$, $T_{e+1,s}(\tau) = T_{e+1,s_0}(\tau)$, and $\forall e' \leqslant e$, T_{e',s_0} has settled down to at least one branching above $T_{e+1}(\tau)$.

If we never redefine $T_{e+1,s}(\tau * 0)(\tau * 1)$ at any $s > s_0$ we are done.

If we never define $T_{e+1,s_1}(\tau * 0)(\tau * 1)$ by Case II at $s_1 > s_0$, then we may only redefine $T_{e+1,s_2}(\tau * 0)(\tau * 1)$ $s_2 > s_0$, by Case III. Since $\forall s > s_0$ the least dummy extension of $T_{e+1,s}(\tau)$ is (τ_1, τ_2), we have $T_{e+1,s_2}(\tau * 0)(\tau * 1) = \tau_1, \tau_2$. This branching cannot become incompatible with $T_{e',s}$ $\forall s > s_2$, $e' < e$, so $\forall s \geqslant s_2$ $T_{e+1,s}(\tau * 0)(\tau * 1) = \tau_1, \tau_2$. So , $\lim_s T_{e+1,s}(\tau * 0)(\tau * 1)\downarrow$.

Lastly suppose some $s_1 \geqslant s_0$ we choose $T_{e+1,s_1}(\tau * 0)(\tau * 1) = \sigma_1, \sigma_2$ by Case II. We claim that $\forall s \geqslant s_1$ $T_{e+1,s}(\tau * 0)(\tau * 1) = \sigma_1, \sigma_2$, and this by Case I. We proceed by proof by contradiction. (Note: Although the proof is detailed it consists of no more than checking that our motivation of boundary strings is accurate).

So suppose some $s > s_1$ $T_{e+1,s}(\tau * 0)(\tau * 1)$ is cancelled. Choose s minimal. (σ_1, σ_2) is cancelled either because one of σ_1, σ_2 has become incompatible with $T_{e',s}$, or a new boundary string for some $e' < e$ has intervened between $T_{e+1,s}(\tau)$ and σ_1 or σ_2. This can come about only by defining $T_{e',s}(\rho * 0)(\rho * 1) = \rho_1, \rho_2$ by Case II,

$e' < e$. (ρ_1, ρ_2) either cause σ_1 or σ_2 to become incompatible (and so $T_{e',s}(\rho) \subset \sigma_1$ or σ_2), e.g.

II-11

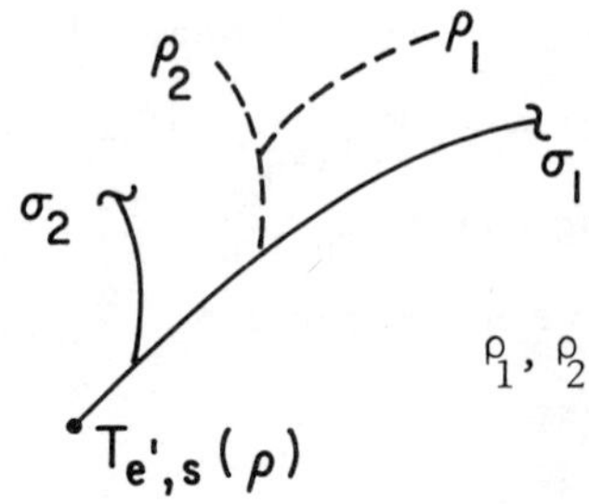

ρ_1, ρ_2 cause σ_1, σ_2 to become incompatible.

or else $T_{e,s}(\tau) \subseteq T_{e',s}(\rho * i) \subset \sigma_j$, $T_{e',s}(\rho * i)$ a new boundary string for e', e.g.

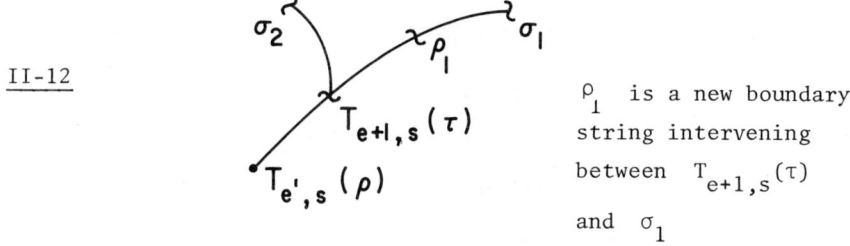

II-12

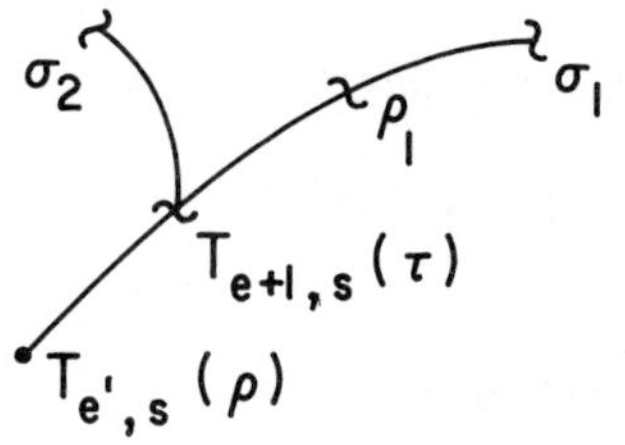

ρ_1 is a new boundary string intervening between $T_{e+1,s}(\tau)$ and σ_1

Choose e' minimal such that such a branching occurs on $T_{e',s}$. Then choose ρ minimal.

Remark: If $T_{e',s}(\rho * 0)(\rho * 1)\downarrow$ by Case III then they can cause no new incompatibility on $T_{e+1,s}$, by the Remark on page 29.

Let $\rho' * i = \rho$, $i = 0$ or 1. Then $T_{e+1,s}(\tau) \subseteq T_{e',s}(\rho) \subset T_{e',s}(\rho * i = \rho)$, since at stage s, $T_{e',s}$ has settled down to at least one branching above $T_e(\tau)$.

II-13

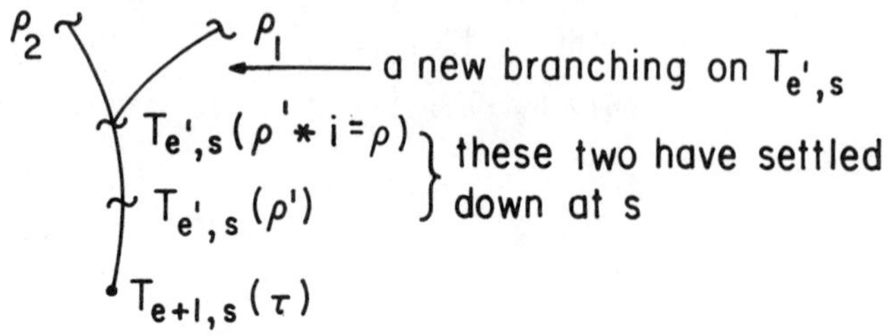

(See page 29 for the proper definition of settling down to one branching above a string, and then convince yourself that we can't have $T_{e',s}(\rho') \subset T_{e+1,s}(\tau)$ due to compatibility).

We claim that $T_{e',s}(\rho) = T_{e',s-1}(\rho)$ is a boundary string for e' at <u>stage s-1</u>. If it is we will be done, since $T_{e+1,s-1}(\tau) \subset T_{e',s-1}(\rho) \subset \sigma_j$ j = 1 or 2 contradicts the choice of $T_{e+1,s-1}(\tau * 0)(\tau * 1) = \sigma_1,\sigma_2$ by Case I or II: $T_{e',s-1}(\rho)$ intervenes.

Can $T_{e',s}(\rho' * 0)(\rho' * 1)\downarrow$ by Case III originally? No. For then why don't we define $T_{e',s}(\rho' * 0)(\rho' * 1) = \rho_1,\rho_2$? It can only be because some δ, $T_{e',s}(\rho') \subset \delta \subset \rho_1$ or ρ_2, δ a boundary string for some $r < e'$. We must have $\delta \subset T_{e',s}(\rho)$ since it doesn't stop us from defining $T_{e',s}(\rho * 0)(\rho * 1) = \rho_1,\rho_2$.

II-14

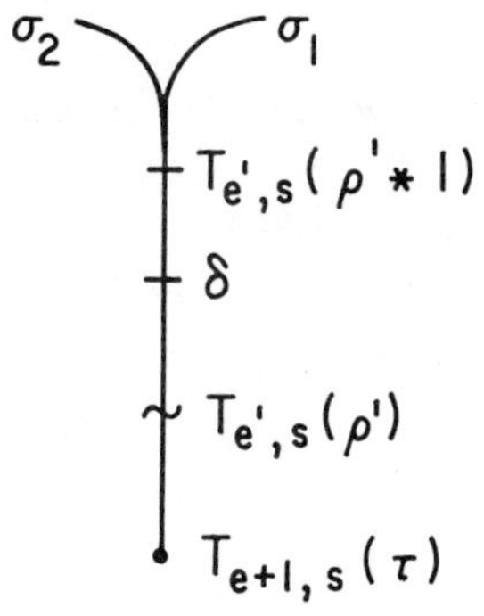

δ could not have been a boundary string for $r < e'$ at stage s-1 or we couldn't have had $T_{e+1,s-1}(\tau * 0)(\tau * 1) = \sigma_1,\sigma_2$ by Case I or II. It couldn't have been newly created at stage s for then e', ρ are not minimal.

So $T_{e',s}(\rho)\downarrow$ Case II originally, $T_{e',s}(\rho) = T_{e',s-1}(\rho)$. Suppose $T_{e',s-1}(\rho * 0)(\rho * 1)\downarrow$ originally by Case II. Why do we redefine $T_{e',s}(\rho * 0)(\rho * 1)$? $T_{e',s-1}(\rho * 0)(\rho * 1)$ must become incompatible at stage s, or a new boundary string intervened. But then, since σ_1, σ_2 are compatible with $T_{e',s-1}$ that incompatibility or new boundary intervening causes σ_1, σ_2 to be cancelled. A contradiction on e', ρ minimal. So $T_{e',s-1}(\rho * 0)(\rho * 1)$ is not defined by Case II if it is defined at all. So $T_{e',s-1}(\rho)$ is a boundary string for e'. Hence $\forall s > s_1\ T_{e+1,s}(\tau * 0)(\tau * 1) = \sigma_1, \sigma_2$. So $\lim_s T_{e+1,s}(\tau * 0)(\tau * 1)$ exists. □

Lemma 2: $B = \lim_s \beta_s$ exists and $\underline{0} < \underline{b} \leqslant \underline{0}'$.

Proof: $\forall e\ T_e(\emptyset)\downarrow$. If $T_{e+1,s}(\emptyset)\downarrow$, $T_{e+1,s}(\emptyset) \supsetneqq T_{e,s}(\emptyset)$. So $T_{e+1}(\emptyset) \supsetneqq T_e(\emptyset)$. All sufficiently large s, $\beta_s \supseteq T_e(\emptyset)$. So $\lim_s \beta_s(x)\downarrow\ \forall x$.

B is not recursive by the choice of $T_{e+1}(\emptyset)$ from $T_e(0)(1)$.

$B \leqslant_T 0'$ since $\{\beta_s\}_{s \geqslant 0}$ is uniformly recursive in s. □

It will follow from $\underline{b}$ minimal that $\underline{b} < \underline{0}'$ as $\underline{0}'$ is not minimal.

Lemma 3: b is minimal.

We will now prune our trees T_e to the $T_e{}^*$ we described in the motivation. Though some care must be taken how we define $T_e{}^*$ and considerable detail must be added to show that our $T_e{}^*$ satisfy the conditions we wish, the $T_e{}^*$ are basically just those of our earlier discussion.

Proof: We construct $T^*_{e,s}$, $T_e{}^* = \lim_s T^*_{e,s}$ satisfying

(1) $T^*_{e,s} \subseteq T^*_{e,s+1}$ and $T_e{}^*$ is partial recursive

(2) T^*_{e+1} is either an e+1 splitting tree

or $\exists\, \beta \subset B$ such that no pair of strings lying above β on T^*_{e+1} e+1-split β.

(3) B lies on $T_e{}^*$

(4) $T^*_{e+1,s}(\delta)$ is a boundary string for $T_{e',s}$ some $e' \leq e+1 \iff T^*_{e+1,s}(\delta)$ is an end string on $T^*_{e+1,s}$

(5) $T^*_{e,s}$ is compatible with $T_{e',s}$ $\forall e' \leq e$

(6) $T_{e,s}(\delta) \subset T_{e,s}(\gamma)$ both lie on $T^*_{e,s} \Rightarrow \forall \rho,\ \delta \subseteq \rho \subseteq \gamma$, $T_{e,s}(\rho)$ lies on $T^*_{e,s}$.

(1), (2) and (3) will yield our lemma to us by an application of the computation lemmas of Chapter I.

(4), (5) and (6) are necessary for our inductive proof of (1), (2) and (3).

The construction of $T_e{}^*$ proceeds:

$$T^*_{0,s} = T_{0,s} \quad \forall s.$$

Assume $\forall r \leq e$, $\forall s$, $T^*_{r,s}$ and $T_r{}^*$ have been defined and satisfy (1) - (6).

There are three cases:

(i) $\exists$ a string π on T_e^* such that $\pi \subset B$ and for no τ, s does $T_{e+1,s}(\tau) \supseteq \pi$.

(ii) B lies on T_{e+1} and $\exists$ a string π on T_e^* such that $\pi \subset B$ and for no τ, s for which $T_{e+1,s}(\tau)\downarrow$ originally by Case II, does $T_{e+1,s}(\tau) \supseteq \pi$ <u>and</u> lie on $T^*_{e,s}$.

(iii) otherwise.

Define

$$\pi(e+1) = \begin{cases} \text{(i)} & \mu\pi \text{ satisfying (i)} \\ \text{(ii)} & \mu\pi \text{ satisfying (ii) and } \pi \supseteq T_{e+1}(\emptyset) \\ \text{(iii)} & \text{the least } T_{e+1}(\tau) \text{ such that } T_{e+1}(\tau) \text{ lies on } T_e^* \text{ and } T_{e+1}(\tau) \subset B, \text{ and } \tau \supsetneq \emptyset. \end{cases}$$

Define $s(e+1) = \mu t(t > s(e)$ and $\forall s > t$, $T_{e',s}(\tau * 1) \cong T_{e',t}(\tau * i)$ $i = 0,1$, for $\forall e' \leqslant e+1$ $\forall \tau$ such that $T_{e'}(\tau)\downarrow \subseteq \pi(e+1))$.

Define, $s > s(e+1)$,

$$T^*_{e+1,s} = \begin{cases} \text{if (i) or (ii) hold, } F_{\pi(e+1)}(T^*_{e,s}) \\ \text{if (iii) holds, "}Sp_{\pi(e+1)}(T_{e+1,s}) \cap T^*_{e,s}\text{"} \end{cases}$$

where we define

$$Sp_{\pi(e+1)}(T_{e+1,s}) \cap T^*_{e,s} = T^*_{e+1,s}$$

by

$$T^*_{e+1,s}(\emptyset) = \pi(e+1)$$

and if

$$T^*_{e+1,s}(\tau)\downarrow = T_{e+1,s}(\delta),$$

let $\rho \supseteq \delta$ be minimal such that $T_{e+1,s}(\rho * 0)(\rho * 1)\downarrow$ by Case II (an e+1-splitting) and <u>both</u> lie on $T^*_{e,s}$.

Set $T^*_{e+1,s}(\tau * 0)(\tau * 1) = T_{e+1,s}(\rho * 0)(\rho * 1)$.

Note: (a) (1) is intended in the functional sense: if $T^*_{e,s}(\delta)\downarrow$, then $T^*_{e,s}(\delta) = T^*_{e,s+1}(\delta)$. Once we show that $\forall s > s(e+1)\ T^*_{e+1,s} \subseteq T^*_{e+1,s+1}$ then $\bigcup_{s>s(e+1)} T^*_{e+1,s} = T^*_{e+1}$ exists, and is clearly partial recursive.

(b) Moreover, it is clear that

if (i) or (ii) hold $\quad T^*_{e+1} = F_{\pi(e+1)}(T^*_e)$

and

if (ii) holds $\quad T^*_{e+1} = Sp_{\pi(e+1)}(T_{e+1}) \cap T^*_e$.

(c) The reader may be curious as to how case (ii) arises. If he will review in the motivation section the explanation of how dummy extensions occur, he will see that it is possible for $T_{e+1,s}$ to keep adding dummy extensions, while never reaching an e+1-splitting. Case I can then keep these on $T_{e+1,s}$.

We proceed by induction, first on e, then on s.

We do not prove (1)-(6) in that order, but rather in the order (5), (6), (4), (1), (3), (2). We will always assume $s > s(e+1)$

First note that $T^*_{0,s} = T_{0,s}$, $T^*_0 = T_0$ satisfy (1), (3), (5) and (6). (1)-(6) hold immediately for $T^*_{1,s}$ and T^*_1. Suppose $e \geqslant 1$.

(5) $T^*_{e+1,s}$ is compatible with $T_{e',s}$ $\forall e' \leq e+1$

Proof: Since $T^*_{e+1,s} \subseteq T^*_{e,s}$, we have that $T^*_{e+1,s}$ is compatible with $T_{e',s}$ $\forall e' \leq e$ by induction on (5). It remains only to show that $T^*_{e+1,s}$ is compatible with $T_{e+1,s}$.

If we define $T_{e+1,s} = Sp_{\pi(e+1)}(T_{e+1,s}) \cap T^*_{e,s}$ then since $T^*_{e+1,s} \subseteq T_{e+1,s}$ we are done.

If we define $T^*_{e+1,s} = F_{\pi(e+1)}(T^*_{e,s})$ by case (i) then every string on $T^*_{e+1,s}$ extends $T_{e+1,s}(\tau) = T_{e+1}(\tau) \subsetneq T_{e+1}(\emptyset)$ an end string on $T_{e+1,s}$. Why? $B \supset T^*_{e+1,s}(\emptyset)$, so some r, $T_{r,s}(\emptyset) = T_r(\emptyset) \supseteq T^*_{e+1}(\emptyset)$ (since $s > s(e+1)$). $T_{r,s}$ is compatible with $T_{e+1,s}$, hence $T_{r,s}(\emptyset)$ extends $T_{e+1,s}(\tau) \subset \pi(e+1)$ an end string on T_{e+1}.

II-15

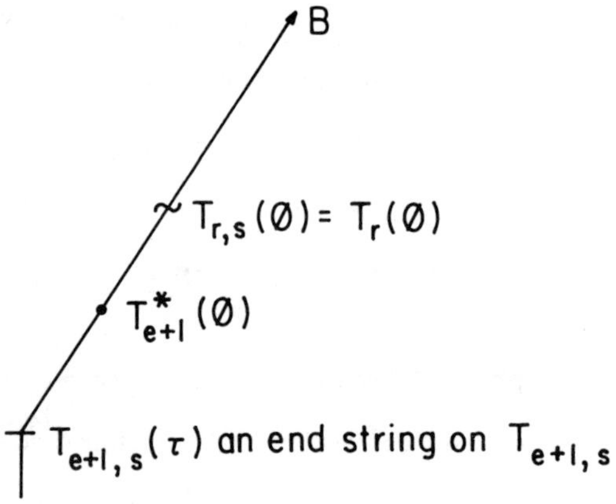

Lastly, suppose $T^*_{e+1,s} = F_{\pi(e+1)}(T^*_{e,s})$ by case (ii). First, $\pi(e+1) = T^*_{e+1,s}(\emptyset)$ lies on $T_{e+1,s}$.

Suppose not.

$$T_{e+1,s}(\gamma * i) = T_{e+1}(\gamma * i)$$

II-16

$$T^*_{e+1,s}(\emptyset) = \pi(e+1)$$

The equalities in this diagram follow from $s > s(e+1)$

$$T_{e+1,s}(\gamma) = T_{e+1}(\gamma)$$

$T_{e+1,s}(\gamma * i)$ $i = 0,1$ must be defined originally by Case II since if it were defined by Case III the minimality condition would force $T_{e+1,s}(\gamma * i) = T^*_{e+1,s}(\emptyset)$ as $T^*_{e+1,s}(\emptyset)$ is compatible with $T_{e',s}$ $\forall e' \leqslant e$. So neither of $T_{e+1,s}(\gamma * 0)(\gamma * 1) = T_{e+1}(\gamma * 0)(\gamma * 1)$ lie on $T_e{}^*$ by definition of case ii.

To arrive at a contradiction we have two cases: First, suppose $\forall\, e' < e$ $T^*_{e'+1} = F_{\pi(e'+1)}(T^*_{e'})$. Then $T_{e+1}(\gamma * 0)(\gamma * 1)$ lie on T^*_e, contradiction. Second, suppose e' is maximal such that $e' \leqslant e$, $T^*_{e'}$ is defined by case (iii). If ρ lies on $T^*_{e'}$ then ρ lies on T^*_e since we have only taken full subtrees since e' (case (i) or (ii) applied to $\forall e''$ $e' < e'' \leqslant e$). The object now is to show that $T_{e+1}(\gamma * i)$ lies on $T^*_{e'}$ for $i = 0$ or 1. $B \supset T_{e+1}(\gamma * i)$ $i = 0$ or 1 since $B \supset \pi(e+1) \supset T_{e+1}(\gamma)$ and B lies on T_{e+1} (as case (iii) applies to $e+1$). $T_{e+1}(\gamma * i)$ $i = 0,1$, cannot both extend end strings on $T_{e'}$ since B lies on $T_{e'}$. So $T_{e+1}(\gamma * i)$ lies on $T_{e'}$

some $i \in \{0,1\}$. As (6) applies to $T^*_{e'}$ and $\pi(e') \subsetneq \pi(e+1)$ we must have $T_{e+1}(\gamma * i)$ lies on $T^*_{e'}$ (see diagram II-17). Hence $T_{e+1}(\gamma * i)$ lies on T^*_e, contradiction.

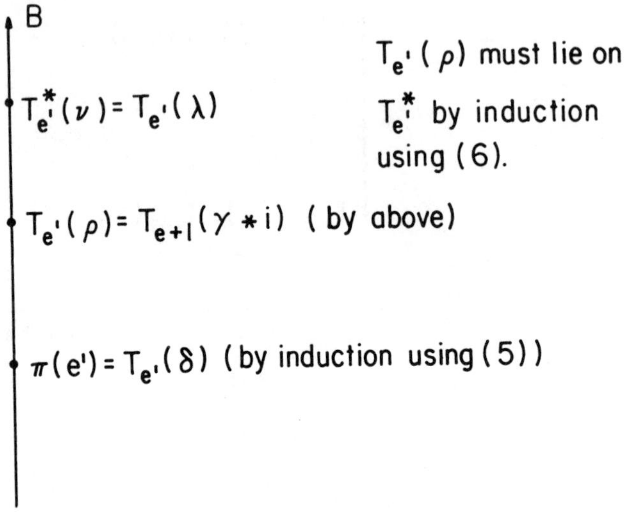

II-17

Hence $\pi(e+1)$ lies on $T_{e+1,s}$ $\forall s > s(e+1)$

Now suppose $T^*_{e+1,s}(\tau)$ lies on $T_{e+1,s}$ and $T^*_{e+1,s}(\tau * 0)(\tau * 1)\downarrow$. Assume that (5) does not hold, and $T^*_{e+1,s}(\tau * 0)(\tau * 1)$ are not compatible with $T_{e+1,s}$. We cannot have diagram II-18,

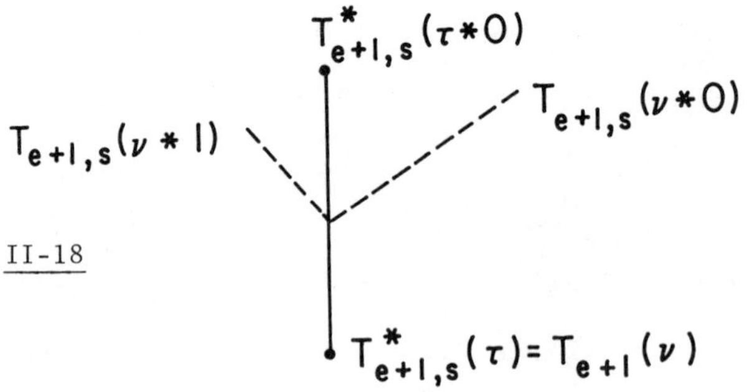

II-18

by dividing into two cases, as above, and in the latter case $T_{e+1,s}(\nu * 0)(\nu * 1)$ must be compatible with $T_{e',s}$. So we must have:

II-19

$$\bullet\, T_{e+1,s}(\nu * i) \quad i = 0 \text{ or } 1$$

$$\bullet\, T^{*}_{e+1,s}(\tau * 0)$$

(I for O is the same)

$$\bullet\, T^{*}_{e+1,s}(\tau) = T_{e+1,s}(\nu)$$

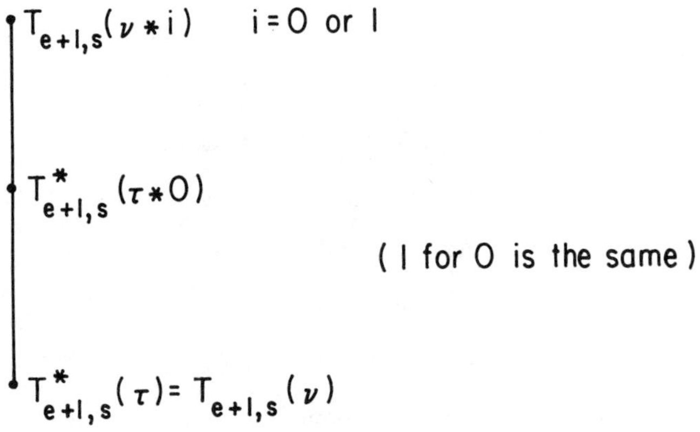

Our argument then proceeds almost as for $\pi(e+1)$. $T_{e+1,s}(\nu * i)$, as above, cannot be defined by Case III originally, by the minimality of such an extension. Hence it was defined by Case II; it may not lie on T^{*}_{e}, as above, by the definition of case (ii) and $\pi(e+1)$. We will show as our contradiction that $T_{e+1,s}(\nu * i)$ must be on T^{*}_{e}. We claim there is a δ such that $T^{*}_{e,s}(\delta) \supset T_{e+1,s}(\nu * i)$ (diagram II-20).

II-20

$$\bullet\, T^{*}_{e,s}(\delta)$$

$$\bullet\, T_{e+1,s}(\nu * i)$$

$$\bullet\, T^{*}_{e+1,s}(\tau * 0)$$

$$\bullet\, T^{*}_{e+1,s}(\tau) = T_{e+1,s}(\nu)$$

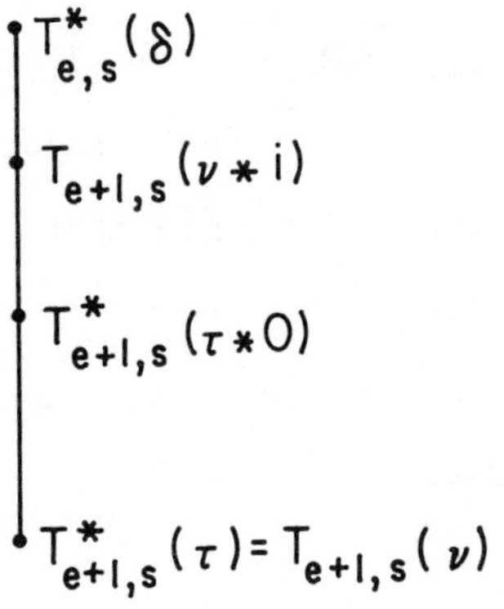

If not, there is a ρ such that $T^*_{e,s}(\rho) \subset T_{e+1,s}(\nu * i)$ and $T^*_{e,s}(\rho)$ is an end string on $T^*_{e,s}$ (since diagram II-18 never happens). By an inductive application of (4), $T^*_{e,s}(\rho)$ must be a boundary string for some $e' \leqslant e$. This contradicts the choice of $T_{e+1,s}(\nu * i)$ by Case II. So such a δ exists.

Now, arguing as we did for $\pi(e+1)$, we must have $T_{e+1,s}(\nu * i)$ lying on $T^*_{e,s}$. For we may divide into two cases, one in which we never took splitting subtrees before which is trivial, and one in which we did, $T^*_{e'}$ being the last. (6) applies to $T^*_{e',s}$ and hence $T_{e+1,s}(\nu * i)$, which lies on $T_{e',s}$ by an argument as for $\pi(e+1)$, lies also on $T^*_{e',s}$. $T_{e+1,s}(\nu * i)$ then lies on $T^*_{e,s}$ too. □

<u>(6)</u> $T_{e+1,s}(\delta) \subset T_{e+1,s}(\gamma)$ both lie on $T^*_{e+1,s}$

$$\Rightarrow \forall \rho,\ \delta \subseteq \rho \subseteq \gamma,$$

$$T_{e+1,s}(\rho) \text{ lies on } T^*_{e+1,s}.$$

Translated, this means that we pick up a "consecutive" subtree of $T_{e+1,s}$ on $T^*_{e+1,s}$.

<u>Proof</u>: If we used case (i) for $T^*_{e+1,s}$ (6) follows because no $T_{e+1,s}(\delta)$ lies on $T^*_{e+1,s}$. If we used case (ii) for $T^*_{e+1,s}$, we may argue as in (5). Indeed, the reader may convince himself that we proved (6) already in that proof.

So we may suppose that we used case (iii) to define $T^*_{e+1,s}$. We will divide into two cases as in (5). First suppose $\forall e' < e$, $T^*_{e'+1} = F_{\pi(e'+1)}(T^*_{e'})$. Then T^*_e is a full subtree of T_0. We encounter no boundary strings on T^*_e by (4) since we have no end strings. So (6) follows.

Now suppose $e' \leqslant e$ is maximal such that $T^*_{e'}$ was defined using case (iii). Then $T^*_{e,s} \subseteq T^*_{e',s} \subseteq T_{e',s}$. Moreover, since (6) holds for $T^*_{e',s}$, and we have only taken full subtrees of $T^*_{e',s}$ since e', if ρ lies on $T^*_{e',s}$ above $T^*_{e,s}(\emptyset)$, it lies on $T^*_{e,s}$.

Now suppose (6) does not hold for $T^*_{e+1,s}$. Then $\exists \delta,\tau,\gamma$ as in diagram II-21.

II-21

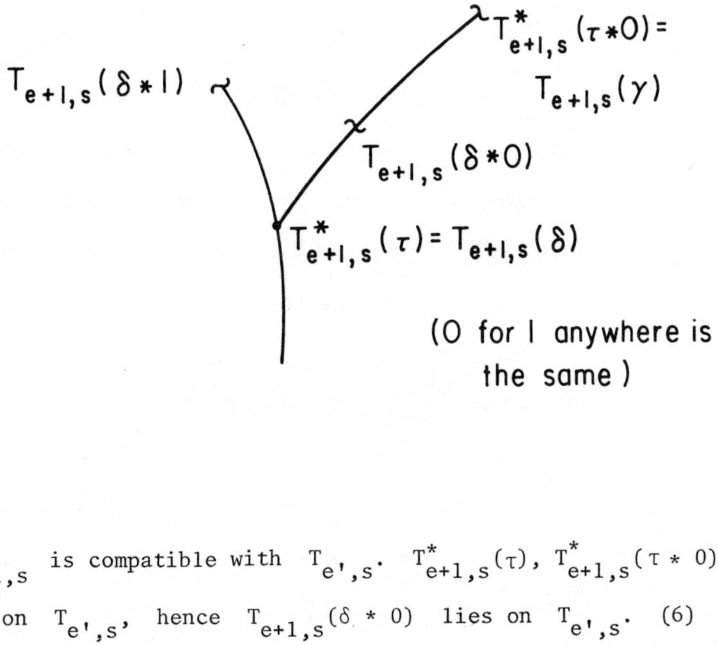

$T_{e+1,s}$ is compatible with $T_{e',s}$. $T^*_{e+1,s}(\tau)$, $T^*_{e+1,s}(\tau * 0)$ lie on $T_{e',s}$, hence $T_{e+1,s}(\delta * 0)$ lies on $T_{e',s}$. (6) applies to $T^*_{e',s}$, hence $T_{e+1,s}(\delta * 0)$ lies on $T^*_{e',s}$. Hence $T_{e+1,s}(\delta * 0)$ lies on $T^*_{e,s}$. We didn't define $T_{e+1,s}(\delta * 0)$ by Case III instead of Case II since then

$T_{e+1,s}(\rho)$ must be a boundary string for $r \leqslant e$, some ρ
$\delta * 0 \subseteq \rho \subset \gamma$

II-22

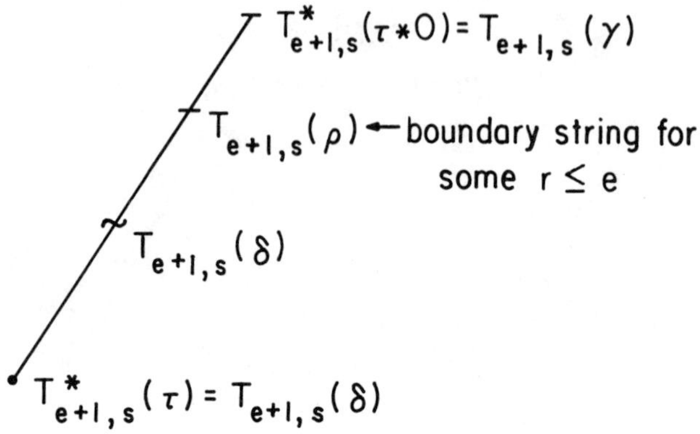

If it weren't a boundary string for some $r \leqslant e$, we would have $T_{e+1,s}(\delta * 0)(\delta * 1)$ redefined as a Case II extension, since there are suitable ones available (e.g. $T^*_{e+1,s}(\tau * 0)(\tau * 1)$). Since a boundary string for $r \leqslant e$ is also an end string on $T^*_{e,s}$, we could not have $T_{e+1,s}(\gamma)$ lying on $T^*_{e+1,s}$ ($T_{e+1,s}(\rho)$ lies on $T^*_{e,s}$ for the same reason $T_{e+1,s}(\delta)$ does). So $T_{e+1,s}(\delta * 0)(\delta * 1)\downarrow$ originally by Case II.

Why do we not place $T_{e+1,s}(\delta * 0)$ on $T^*_{e+1,s}$? It can only be because $T_{e+1,s}(\delta * 1)$ does not lie on $T^*_{e,s}$. But then $T_{e+1,s}(\delta * 1)$ must extend an end string on $T^*_{e,s}$:

II-23

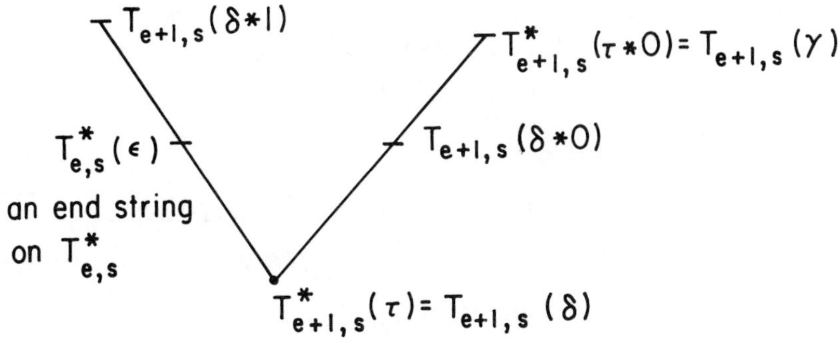

Why? Because $T_{e+1,s}(\delta * 1)$ is compatible with $T_{r,s}$ $\forall r \leq e$. It either lies on $T_{r,s}$ or extends an end string on $T_{r,s}$. If it lay on $T_{r,s}$ $\forall r \leq e$ the only way it could not appear on $T^*_{e,s}$ is if a boundary string, hence an end string on $T^*_{e,s}$ preceded it. If it doesn't lie on $T_{r,s}$ some $r \leq e$, it extends an end string on $T_{r,s}$ hence on $T^*_{r,s}$. $T^*_{e,s}(\varepsilon)$, being an end string on $T^*_{e,s}$ is a boundary string for $T_{r,s}$ for some $r \leq e$, by induction on (4). This contradicts the choice of $T_{e+1,s}(\delta * 0)(\delta * 1)$ by Case II. □

(4) $T^*_{e+1,s}(\delta)$ is a boundary string for $T_{e',s}$ some $e' \leq e + 1$

$\Longleftrightarrow$ $T^*_{e+1,s}(\delta)$ is an end string on $T^*_{e+1,s}$.

Proof: => Since (4) holds for $T^*_{e,s}$ we need only concern ourselves with boundary strings on $T_{e+1,s}$ lying on

$T^*_{e+1,s}$.

If $T^*_{e+1,s}$ is defined by case (i) no string lying on $T_{e+1,s}$ lies on $T^*_{e+1,s}$, so we are done.

If $T^*_{e+1,s}$ is defined by case (ii) no string $T_{e+1,s}(\tau)\downarrow$ by Case II lies on $T^*_{e+1,s}$ so we are done.

Suppose $T^*_{e+1,s} = Sp_{\pi(e+1)}(T_{e+1,s}) \cap T^*_{e,s}$. Can we have $T_{e+1,s}(\tau) = T^*_{e+1,s}(\rho)$, $T_{e+1,s}(\tau * 0)(\tau * 1)\downarrow$ by Case III, and yet $T^*_{e+1,s}(\rho)$ not an end string on $T^*_{e+1,s}$? No. For then we would have:

II-24

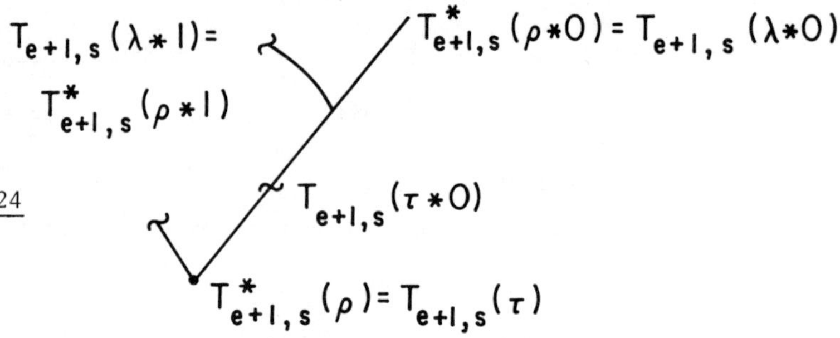

and by (6) we would then have to have $T_{e+1,s}(\tau * 0)$ on $T^*_{e+1,s}$. That contradicts the definition of $T^*_{e+1,s}$.

$\Leftarrow$ If $T^*_{e+1,s} = F_{\pi(e+1)}(T^*_{e,s})$ the result follows by induction on (4).

So suppose we used case (iii) to define $T^*_{e+1,s}$ and (4) fails in this direction. Then $\exists\tau$,

$T^*_{e+1,s}(\tau)$ is an end string for $T^*_{e+1,s}$ but not for $T^*_{e,s}$.

$T^*_{e+1,s}(\tau) = T_{e+1,s}(\gamma)$ some γ. For (4) to fail we need $T_{e+1,s}(\gamma * 0)(\gamma * 1)$ defined by Case II, but $T_{e+1,s}(\gamma * 1)$, say, does not lie on $T^*_{e,s}$.

Now, as in (6), we can argue that if $T_{e+1,s}(\gamma * 1)$ does not lie on $T^*_{e,s}$, it extends an end string on $T^*_{e,s}$. Hence, by induction on (4) it extends a boundary string for some $e' \leqslant e$, a contradiction on the choice of $T_{e+1,s}(\gamma * 0)(\gamma * 1)$ by Case II. □

(1) $T^*_{e+1,s} \subseteq T^*_{e+1,s+1}$ and T^*_{e+1} is partial recursive.

Proof: By this we mean if $T^*_{e+1,s}(\delta)\downarrow$, $T^*_{e+1,s}(\delta) = T^*_{e+1,s+1}(\delta)$.

If $T^*_{e+1,s}$ is defined by case (i) or (ii) (1) follows by induction.

So suppose $T^*_{e+1,s}$ is defined by case (iii).

$$T^*_{e+1,s}(\emptyset) = T^*_{e+1,s+1}(\emptyset) \text{ by definition.}$$

Suppose $\tau \supset \emptyset$ and

$$T_{e+1,s}(\rho) = T^*_{e+1,s}(\tau); \text{ then}$$

we need
$$T_{e+1,s+1}(\rho) = T_{e+1,s}(\rho) = T^*_{e+1,s+1}(\tau).$$

If we can show $T_{e+1,s+1}(\rho) = T_{e+1,s}(\rho)$ the last equality will follow, for $T^*_{e,s} \subseteq T^*_{e,s+1}$ and by induction on $lh(\tau)$, $T_{e+1,s}$ has settled down below $T_{e+1,s}(\rho)$ on this branch.

$T_{e+1,s}(\rho)\downarrow$ originally by Case II. It lies on $T^*_{e,s}$ and hence on $T^*_{e,s+1}$, so by (5) it is compatible with $T_{e',s+1}$ $\forall e' \leqslant e$. If no new boundary strings intervene we must have, by Case I, $T_{e+1,s+1}(\rho' * 0)(\rho' * 1) = T_{e,s}(\rho' * 0)(\rho' * 1)$ where $\rho' * i = \rho$, $i \in \{0,1\}$, (all the above applies to $\rho' * 1\text{-}i$ too). So we must only show that no new boundary string intervenes.

Suppose one does. That is γ, a new boundary string for $e' \leqslant e$, occurs at stage $s+1$, and $T_{e+1,s}(\rho') \subseteq \gamma \subset T_{e+1,s+1}(\rho)$.

II-25

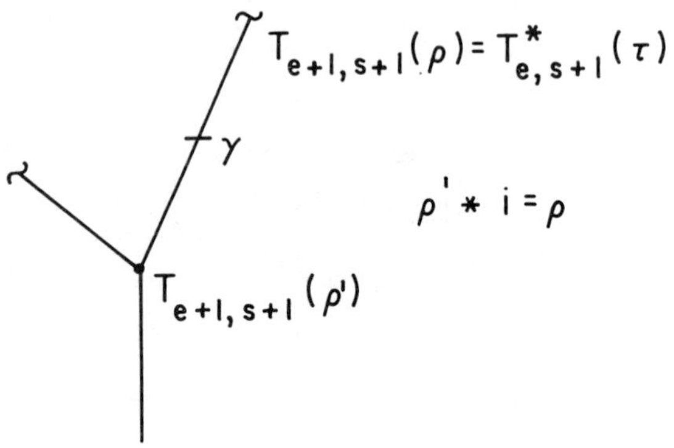

Now argue as in (5) that γ must lie on $T^*_{e,s+1}$. So γ is an end string on $T^*_{e,s+1}$ by induction on (4). But that contradicts $T_{e+1,s+1}(\rho)$ lying on $T^*_{e,s+1}$.

T^*_{e+1} is partial recursive by note a following the definition of $T^*_{e+1,s}$. □

Note: The reader should be aware of how we suppressed the bad effects of infinite injury to achieve (1). (1) occurs because we never have to throw away splittings of $T_{e+1,s}$ lying on $T^*_{e+1,s}$. And this is because $T^*_{e,s}$ not only controls all new incompatibilities caused by $T_{e'}$ $e' \leqslant e$, it even controls the ghosts of incompatibility, not allowing new boundary strings to intervene.

<u>(3)</u> B lies on T^*_{e+1}.

<u>Proof</u>: If we used case (i) or (ii) to define T^*_{e+1},

(3) follows by induction and the definition of $\pi(e+1)$.

So suppose we used case (iii) to define T^*_{e+1}. Suppose (3) fails for T^*_{e+1}. B is compatible with T_r, $\forall r$. Hence there is an end string of T^*_{e+1} such that $T^*_{e+1}(\tau) \subset B$, as $T^*_{e+1} \subseteq T_{e+1}$.

Since $T^*_{e+1}(\tau)$ is a permanent end string on T^*_{e+1}, by (4) it must be a permanent boundary string for some $e' \leqslant e + 1$. If it were a boundary string for $e' \leqslant e$, it would be a permanent end string on T^*_e by (4). But then T^*_e would not satisfy (1). Hence $T^*_{e+1}(\tau)$ is a boundary string only for $e + 1$.

Let $T^*_{e+1}(\tau) = T_{e+1}(\delta)$. Since we are in case (iii), we know that $\exists \gamma \supsetneqq \delta$ such that $T_{e+1}(\gamma * 0)(\gamma * 1){\downarrow}$ by Case II and $T_{e+1}(\gamma * 0)$, say, lies on T^*_e. $\gamma \neq \delta$ since $T_{e+1}(\delta)$ is a boundary string for $e + 1$. Choose the least such γ.

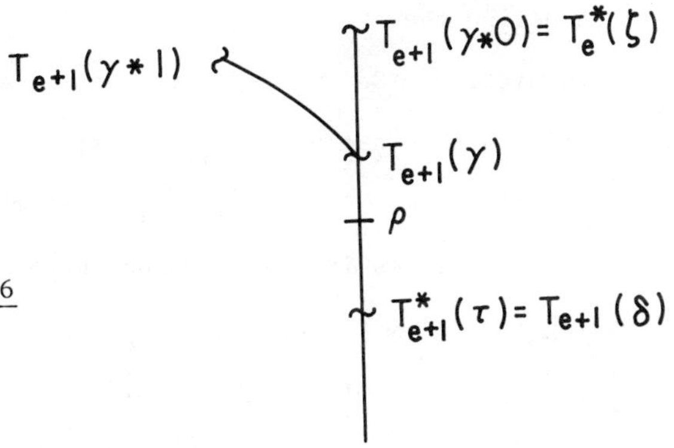

<u>II-26</u>

But then some $\rho \subset T_{e+1}(\gamma)$ is a boundary string for some $e' \leqslant e$. If not we would define $T_{e+1}(\delta * 0)(\delta * 1)$ as a splitting pair.

Now proceed as in (5) to find that ρ lies on T^*_e. Hence ρ is an end string on T^*_e, contradicting that $T_{e+1}(\gamma * 0)$ lies on T^*_e. $\square$

(2) T^*_{e+1} is either an e+1-splitting tree

or $\exists\, \beta \subset B$ such that no pair of strings lying on T^*_{e+1} e+1-split β.

Proof: If T^*_{e+1} is defined by case (iii) it is immediate.

So suppose we defined T^*_{e+1} by case (i), and (2) fails. Then $\exists \tau_1, \tau_2$ such that $T^*_{e+1}(\tau_1)$, $T^*_{e+1}(\tau_2)$ e+1-split $T^*_{e+1}(\emptyset)$. As we deduced in (5), $T^*_{e+1}(\emptyset) \supseteq T_{e+1}(\tau)$ an end string on T_{e+1}. $T^*_{e+1}(\tau_1)$, $T^*_{e+1}(\tau_2)$ are compatible with $T_{e'}$ $\forall\, e' \leqslant e$ by (5). So $\forall s > s_0$ Case II holds for $T_{e+1,s}(\tau * 0)(\tau * 1)$, with the exception possibly of II b). Hence $T_{e+1}(\tau * 0)(\tau * 1)$ must be defined, as we must define them by Case II or Case III at $\forall s > s_0$. Contradiction. So $T^*_{e+1}(\emptyset)$ is the β we wish.

Suppose we defined T^*_{e+1} by (ii) and (2) fails. Take τ_1, τ_2 as above. $T^*_{e+1}(\tau_1)$, $T^*_{e+1}(\tau_2)$ are compatible with $T_{e'}$ $\forall\, e' \leqslant e$. As in (6) we cannot have ρ, a boundary string for $e' \leqslant e$ such that $T^*_{e+1}(\emptyset) = T_{e+1}(\gamma) \subseteq \rho \subsetneq T^*_{e+1}(\tau_j)$ j = 1 or 2. (Divide into two cases; in the case where we have $T^*_{e'}$ defined by case (iii), ρ would lie

on $T^*_{e'}$, hence on T^*_{e+1} and so be an end string on T^*_{e+1}).
Therefore, we should eventually define

$$T_{e+1}(\gamma * 0)(\gamma * 1) = T^*_{e+1}(\tau_1),\ T^*_{e+1}(\tau_2)$$

by Case II. This we never do since case (ii) holds.
Contradiction. So $T^*_{e+1}(\emptyset)$ is the β we wish. □

This concludes Lemma 3 and the proof of the theorem.

CHAPTER III: A MINIMAL DEGREE such that $\underline{m}' = \underline{0}'$

Sections

* * *

The existence of such a degree was first proved by Yates (13). It was a corollary to his main theorem, which we prove in Chapter V, that given $\underline{a}$ r.e., $\underline{a} > \underline{0}$, there is a minimal degree $\underline{m} < \underline{a}$.

We give a direct proof of $\underline{m}' = \underline{0}'$ for two reasons. First, the proof in Chapter IV that given $\underline{c} \geqslant \underline{0}'$ we can find a minimal degree such that $\underline{m}' = \underline{c}$ is an expansion of this construction. We believe it is easiest to understand it in that perspective. That theorem was proved by Cooper (2) and it is there that he first developed the technique of adding "followers" to a construction. That technique is useful in any situation in which a finite injury argument needs to be added to an $\underline{m} < \underline{0}'$ construction. $\underline{m}' = \underline{0}'$ is a particularly easy application of "followers" and that is the second reason for this Chapter.

Motivation

We first introduce the jump operator:

$$J(A,x) = \begin{cases} \Phi_x(A,x) + 1 & \text{if } \Phi_x(A,x)\downarrow \\ 0 & \text{otherwise} \end{cases}$$

We call $J(A)$ the "jump of A" and denote it by A'.

$$J_s(A,x) = \begin{cases} \Phi_{x,s}(A,x) + 1 & \text{if defined} \\ 0 & \text{otherwise} \end{cases}$$

Then $\lim_s J_s(A,x) = J(A,x)$ and $J_s(\rho,x)$ similarly.

$J_s(A,x)$ is recursive in s and A.

If $J_s(\gamma,x) > 0$, then $\forall \rho \supset \gamma,\ t > s,\ J_t(\rho,x) = J_s(\gamma,x)$.

* * *

If we construct a minimal degree $\underline{m} < \underline{0}'$ we immediately have $\underline{0}' \leqslant \underline{m}'$. We must somehow push $\underline{m}'$ down. If we could arrange for our approximation β_s of the previous chapter to also satisfy $\lim_s J_s(\beta_s,e) = J(B,e)$, then we would have $J(B,e) \leqslant_T 0'$ by Shoenfield's lemma. This is what we do.

* * *

Let's fix e. Suppose by stage $s_0, T_{e,s_0}(\emptyset)(0)(1)$ and $\phi_{e,s_0}(x)$ $\forall x < lh(T_{e,s_0}(i))$ $i = 0,1$ have settled down. As before we'll have $\beta_s \supseteq T_e(i)$, $\forall s > s_0$, where $T_e(i) \not\subseteq \phi_e$. For convenience suppose $i = 0$.

If we never have $J_s(\beta_s,e) > 0$, $s > s_0$, then we know that $J(B,e) = 0 = J_s(\beta_s,e)$ $\forall s > s_0$. So $J_{s_0}(\beta_{s_0},e)$ has settled down. So suppose at $s_1 > s_0$, s_1 minimal, $J_{s_1}(\beta_{s_1},e) > 0$.

III-1

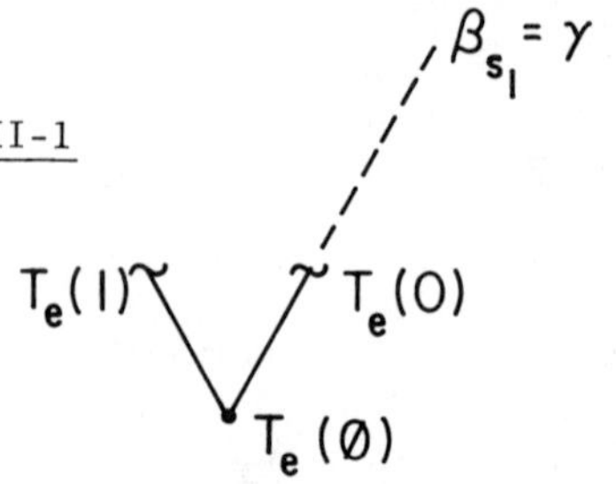

We would like for B to pass through this γ. For if it did then $J(B,e) = J_s(\gamma,e) = J_s(\beta_s,e)$ all sufficiently large s.

So we will set $\gamma = \beta_{s_1}$ <u>as a follower of $T_e(0)$</u>, $\forall s > s_1$. And then we will require $T_{e+1,s}(\emptyset) \supseteq \gamma$, $\forall s > s_1$.

But wait, further branching on $T_{e',s}$, $e' < e$, could cause B to branch away from γ, viz:

III-2

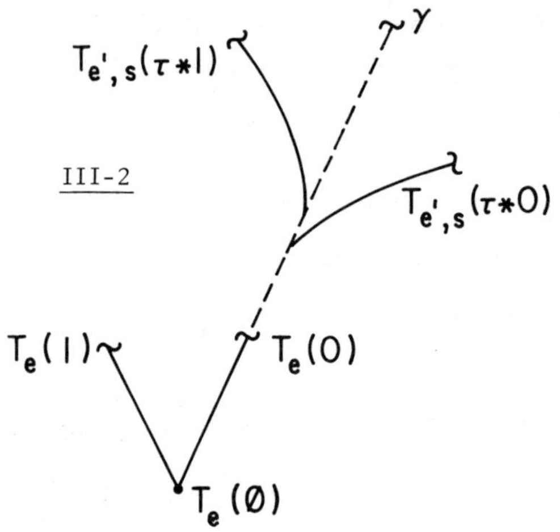

Then we could not put $T_{e+1,s}(\emptyset)$ above γ and still have $T_{e+1,s}(\emptyset)$ compatible with $T_{e',s}$.

For $T_{e',s}$ to force us away from γ as in diagram III-2 $T_{e',s}(\tau) \supseteq T_e(0)$, since $T_e(0)$ has settled down and is compatible with $T_{e',s}$.

To protect γ from this we will simply prohibit such branching. That is, with γ we associate the <u>prohibition</u>:

> every σ incompatible with γ is prohibited as an extension of any δ, $T_e(0) \subseteq \delta \subset \gamma$, at any stage $s > s_1$.

(It is only necessary to apply this prohibition to new Case II branchings. Case III branchings, being dummy ones, cannot force B away from γ).

Thus the only way we can cancel γ is to cancel $T_{e,s_1}(0)$. That is, γ has the same priority as $T_{e,s_1}(0)$.

We will now be able to put $T_{e+1}(\emptyset)$ above γ, and so B will pass through γ. Hence $J_{s_1}(\beta_{s_1},e)$ has settled down.

But how do we know, if there are e'-splittings lying on $T^*_{e'-1}$ above every beginning of B, that we will always be able to capture e'-splittings on $T_{e'}$? Won't prohibitions interfere? Prohibitions do interfere, but only in a nice finite manner. Consider diagram III-3.

III-3

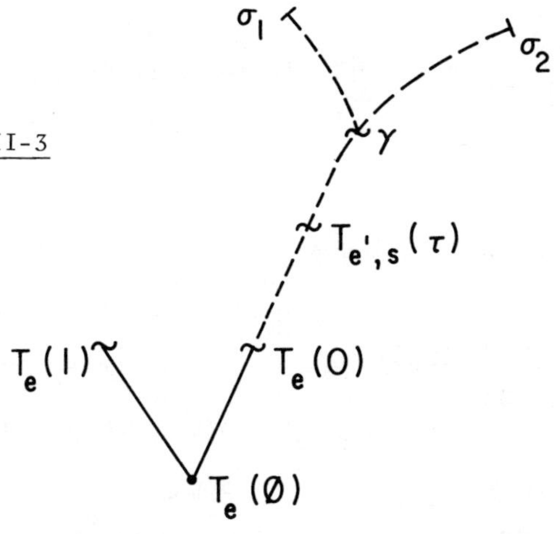

Suppose $T_{e',s}(\tau) = T_{e'}(\tau)$ and is an end string on $T_{e'}$. If there really are e'-splittings lying on $T^*_{e'-1}$ above every beginning of B, there must be one lying above γ, say (σ_1,σ_2). σ_1,σ_2 are not prohibited as extensions of $T_{e',s}(\tau)$ since both are compatible with γ. So we may set $T_{e',s}(\tau * 0)(\tau * 1) = \sigma_1,\sigma_2 = T_{e'}(\tau * 0)(\tau * 1)$.

III-4

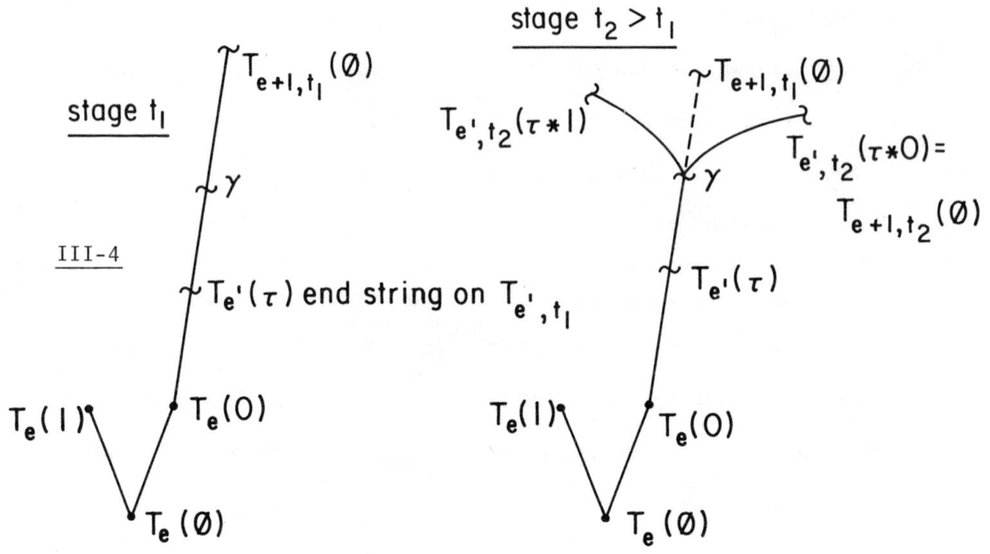

Any δ following $T_{e+1,t_1}(0)$, say, will be knocked off at stage t_2 by T_{e',t_2}.

If we erect another permanent follower δ above γ, viz:

III-5

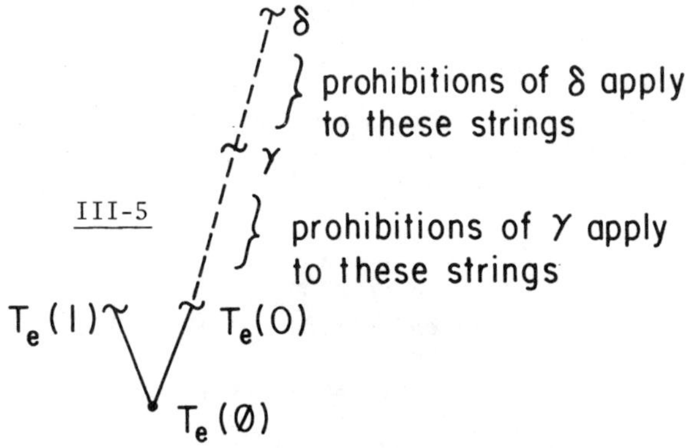

then δ must follow some $T_m(i) \supseteq \gamma$ $(m > e,\ i \in \{0,1\})$. So δ's prohibitions will not apply to strings below γ.

That is, we do not get the effect of an infinite follower.

It is important to realize that our prohibitions impede branching on $T_{e'}$, $e' \leqslant e$, but cause no cancellations. This will allow us to carry over Lemma 1, that $\lim_s T_{e,s}$ exists, from Chapter II, except for the proof that $\lim_s T_{e,s}(\emptyset)\downarrow$.

* * *

It can now be seen that followers are useful in any situation in which we wish to attach to a minimal degree a condition which for every e can be fulfilled by a single string. If we view followers as keeping us from using the wrong strings, e.g. ones incompatible with γ as in III-2, then other uses become possible.

Note: In Cooper's first full approximation construction (1) he used a technique similar to followers. He simply "stretched" $T_{e',s}(\tau)$ in diagram III-3 to $T_{e',s+1}(\tau) = \gamma$. The effect is the same. Having no infinite follower is equivalent to $T_{e',s}(\tau)$ not being stretched infinitely often. However, stretching enormously complicates the proof of Lemma 3 of this Chapter. In (2) Cooper switched to followers which we find much easier.

Construction of $\underline{m}' = \underline{0}'$

A minimal degree such that $\underline{m}' = \underline{0}'$

Since this construction is so similar to that of $\underline{m} < \underline{0}'$ we will only list the changes from that one.

$T_{e,s}, e > 0$

$T_{e,s}(\emptyset)$: let i be minimal, $i \in \{0,1\}$, such that

$T_{e-1,s}(i) \not\subseteq \phi_{e,s}$

If there is a follower for $e - 1$, γ, of $T_{e-1,s}(i)$ set $T_{e,s}(\emptyset)$ = the least $\delta \supseteq \gamma$ such that δ is compatible with $T_{e',s}$ $\forall e' < e$.

Otherwise set $T_{e,s}(\emptyset) = T_{e-1,s}(i)$.

Case II: add (d) σ_1, σ_2 are not prohibited at stage s as extensions of $T_{e,s}(\tau)$.

add: at the end of stage s we proceed to

Appointment of followers and prohibitions, cancellations.

Let $e > o$ be minimal such that e does not have a follower, $T_{e,s}(0)(1)\downarrow$, and $J_s(\beta_s, e) > 0$.

If no such e exists, go on to stage $s + 1$.

If there is such an e then if $\beta_s \supset T_{e,s}(i)$ $i = 0$ or 1, set $\gamma = \beta_s$ a follower for e of $T_{e,s}(i)$.

If at any stage $t + 1 > s$

$$T_{e,t+1}(0)(1) \neq T_{e,t}(0)(1)$$

or

$$\beta_t \not\supset T_{e,t}(i), \quad i \text{ as above,}$$

then immediately cancel γ as a follower, along with its associated prohibitions.

We prohibit any string σ, $\sigma|\gamma$ as an extension of any ρ such that $T_{e+1,t}(i) \subseteq \rho \subset \gamma$, i as above, at all stages $t > s$ at which γ is a follower for e.

* * *

This ends the construction.

Proof

Lemma 1: $\lim_s T_{e,s}(\tau)$ exists $\forall e,\tau$.

Proof: First note that, except for the beginnings of trees, $T_{e,s}(\emptyset)$, no branchings are ever cancelled because of followers or prohibitions. We impede the growth of splittings, but we do not cancel them.

Thus $\lim_s T_{1,s}(\tau)$ exists $\forall\tau$ follows as before, as no branch is ever cancelled on this tree.

$\lim_s T_{e+1,s}(\emptyset)\downarrow$: Suppose we have $\lim_s T_{e',s}(\tau)$ exists $\forall e' \leq e$, $\forall\tau$.

Let s_0 be such that $\forall s > s_0$ $T_{e,s}(0)(1) = T_{e,s_0}(0)(1)$ and $\forall x \leq lh(T_{e,s_0}(0)(1))$, $\phi_{e+1}(x) \cong \phi_{e+1,s_0}(x)$. Then $\forall s > s_0$ $T_{e+1,s}(\emptyset) \supseteq T_{e,s}(i)$, $i \in \{0,1\}$, i fixed.

If at no stage $s_1 > s_0$ does e have a follower then $\forall s > s_0$ $T_{e+1,s}(\emptyset) = T_{e,s}(i)$.

So suppose there is a stage $s_1 > s_0$ at which we appoint a follower γ of $T_{e,s}(i)$ for e.

Let $s_2 > s_1$ be such that the least $\delta \supseteq \gamma$ such that δ is compatible with $T_{e'}$ $\forall e' \leq e$ is well defined. Such a stage exists. Suppose it did not. Then, considering our argument in Lemma 1 of $\underline{m} < \underline{0}'$, we must have a least $s > s_1$ such that $T_{e',s}(\rho) \subsetneq \gamma$ and γ is not compatible with $T_{e',s}(\rho * 0)(\rho * 1)$ (diagram III-6). $T_{e',s}(\rho * 0)(\rho * 1)$ must be defined by Case II, for Case III branchings cannot cause new incompatibility (this is why prohibitions need only apply to new Case II extensions). But then, because γ's prohibitions are in effect at stage s, $T_{e',s}(\rho) \subsetneq T_{e,s}(i)$.

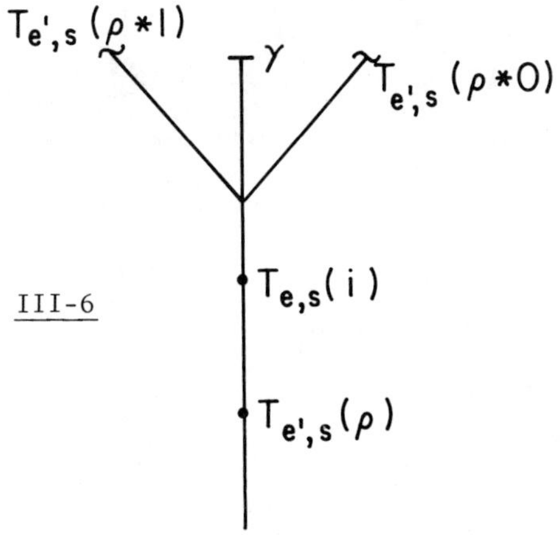

III-6

But then $T_{e,s}(i)$ becomes incompatible with $T_{e',s}$, a contradiction on $s > s_0$.

Hence s_2 and δ exist, and $\forall s \geq s_2\ T_{e+1,s}(\emptyset) = \delta$, So $\lim_s T_{e+1,s}(\emptyset)\downarrow$.

Note: This is a formal proof that <u>we can only knock off γ</u>, a follower of $T_{e,s}(i)$, <u>by knocking off $T_{e,s}(i)$ as well</u>.

The proof that $\lim_s T_{e,s}(\tau)$ exists for $\tau \supsetneq \emptyset$ is exactly as in Lemma 1 of $\underline{m} < \underline{0}'$, for as we remarked, Case III continues to be the same. Appointing γ a follower for e cannot cause cancellation of branches on $T_{e',s}$, $e' < e$. We may impede branching with followers, but we do not cancel branchings due to them. $\square$

<u>Lemma 2</u>: (i) $\lim_s \beta_s = B$ exists

(ii) $\underline{0} < \underline{b} \leq \underline{0}'$

(iii) $\underline{b}' = \underline{0}'$

<u>Proof</u>: (i) and (ii) are exactly as in Lemma 2 of $\underline{m}' < \underline{0}'$.

(iii) Given $e > 0$. Let s_0 be such that $T_{e,s_0}(0)(1)$ have settled down and $\forall s > s_0\ \beta_s \supseteq T_{e,s}(i)$, i fixed $\in \{0,1\}$.

Suppose $\forall s > s_0\ J_s(\beta_s,e) = 0$. Then $\lim_s J_s(\beta_s,e) = 0$.

So suppose some $s_1 > s_0$, $J_{s_1}(\beta_{s_1},e) > 0$. Then if e does not have a follower we will appoint one at stage s_1. If γ follows $T_{e,s_1}(i)$ at stage s_1, it will continue to do so at stage s $\forall s > s_1$, for all the conditions to retain it are in effect. We will have $T_{e+1,s}(\emptyset)\downarrow \supseteq \gamma$ $\forall s > s_1$. Hence $\forall s > s_1$, $\beta_s \supseteq \gamma$ and $J_s(\beta_s,e) = J_{s_1}(\gamma,e)$. So $\lim_s J_s(\beta_s,e) = J(\gamma,e)$. Since $B \supset \gamma$, $J(B,e) = J(\gamma,e)$. Thus $J(B,e) = \lim_s J_s(\beta_s,e)$, $\forall e > 0$. Therefore, $J(B,e) \leqslant_T 0'$. □

Lemma 3: $\underline{b}$ is minimal.

Proof: The T^*_e are constructed exactly as in Lemma 1 of $\underline{m} < \underline{0}'$. The proofs of (1)-(6) are the same, also, with the exception of (2) which we prove here.

(2) T^*_{e+1} is either an e+1-splitting tree or $\exists \beta \subset B$ such that no pair of strings lying on T^*_{e+1} e+1-split β.

Proof: If we define T^*_{e+1} by case (iii) it is immediate.

Suppose we define T^*_{e+1} by case (i).

We have an end string $T_{e+1}(\tau) \subset T^*_{e+1}(\emptyset)$. Since $B \supset T^*_{e+1}(\emptyset)$ we may find a maximal r such that $T_r(i) \subseteq T_{e+1}(\tau)$, $i = 0$ or 1. Let s_0 be such that $\forall s > s_0$ $T_{r,s}(i) = T_{r,s_0}(i)$. If $T_{r,s}(i)$ has no follower, $\forall s > s_0$, proceed as in (2) of Lemma 3 of $\underline{m} < \underline{0}'$ (page 50). If at stage $s_1 \geqslant s_0$ $T_{r,s_1}(i)$ has a

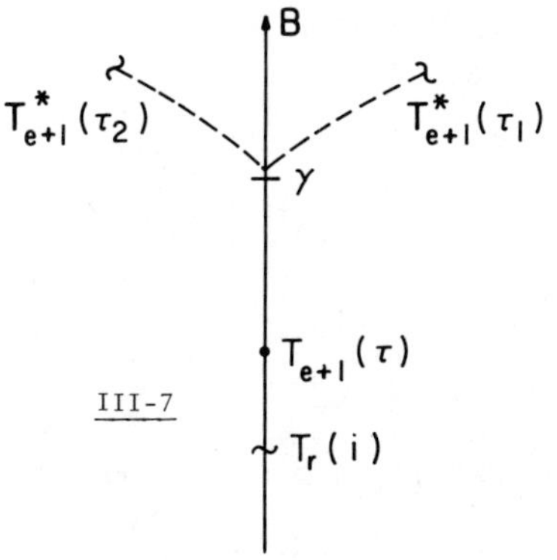

III-7

follower γ, then γ is the follower of $T_{r,s}(i)$ $\forall s > s_1$. Since $\gamma \subset B$, $T_{e+1}(\tau)$ and γ are compatible. At $\forall s > s_1$, no string $\sigma \supset \gamma$ is prohibited as an extension of $T_{e+1}(\tau)$, (if it were it would be prohibited by δ a follower of $T_{m,s}(i) \subseteq T_{e+1}(\tau)$, $m > r$, contradiction).

Suppose $\exists \tau_1, \tau_2$ such that $T^*_{e+1}(\tau_1)$, $T^*_{e+1}(\tau_2) \supset \gamma$ and e+1-split. Since they are compatible with $T_{e'}$ $\forall e' \leqslant e$, and are not prohibited as extensions of $T_{e+1}(\tau)$, we must define $T_{e+1,s}(\tau * 0)(\tau * 1)$ by Case II or Case III at $\forall s > s_2$, some $s_2 \geqslant s_1$. So we must define $T_{e+1}(\tau * 0)(\tau * 1)$, contradiction. Hence the β we wish is $\beta = \gamma$.

Suppose we define T^*_{e+1} by <u>case (ii)</u>. Then let γ be the longest permanent follower of any $T_r(i) \subseteq \pi(e+1)$, $i = 0,1$. Then proceed as in (2) of Lemma 3 of $\underline{m} < \underline{0}'$ (page 50), to see that no e+1-splittings lie on T^*_{e+1} above γ. Hence the β we wish is $\beta = \gamma$. □

This completes the proof of the theorem.

CHAPTER IV: MINIMAL DEGREES AND THE JUMP OPERATOR

Sections

* * *

In this Chapter we describe how to prove that if $\underline{c} \geqslant \underline{0}'$ there is a minimal degree such that $\underline{m}' = \underline{c}$. As we mentioned before, this was proved by Cooper (2). A complete construction and proof is given in that paper (Theorem 1). We content ourselves with providing a fairly detailed description of how to modify the construction of $\underline{m}' = \underline{0}'$ to achieve this theorem. Since the modification is intuitive, and we believe reasonably straightforward, and since the proofs of the necessary lemmas are almost identical to corresponding ones in Chapter III we see no reason to proceed further than a description here.

Friedberg (3) showed that if $\underline{c} \geqslant \underline{0}'$ then there is some $\underline{a}$ such that $\underline{a}' = \underline{c}$. Cooper's theorem shows that this equation is solvable with $\underline{a}$ minimal. In some sense this shows that minimal degrees are spread throughout the degrees.

Shoenfield (10) showed that if $\underline{0}' \leqslant \underline{c} \leqslant \underline{0}''$ and $\underline{c}$ is r.e. in $\underline{0}'$ then some $\underline{a} \leqslant \underline{0}'$, $\underline{a}' = \underline{c}$. Under "Further Topics" we discuss the work that has been done on trying to replace $\underline{a}$ with $\underline{m}$ minimal in this equation.

Motivation

Let $\underline{c} \geq \underline{0}'$, $C \in \underline{c}$. Take $\underline{a}$ such that $\underline{a}' = \underline{c}$ and C^s an $\underline{a}$ recursive approximation of C (i.e. $\lim_s C^s(x) = C(x)$).

We already have, from Chapter III, a minimal degree $\underline{m}$ such that $\underline{m}' = \underline{0}'$. So we will assume $\underline{c} > \underline{0}'$. This allows us to dispense with diagonalizing against ϕ_e, $e > 0$, as $\underline{m}' = \underline{c} > \underline{0}' \Rightarrow \underline{m} \neq \underline{0}$.

Let's ask how we can construct a set, B, with prescribed jump $\underline{c}$. To push the jump up we want C to be Δ_2^0 in B.

> Note: C is Δ_2^0 in B if C can be expressed as $\exists x \forall y P$, and also as $\forall x \exists y Q$ where P and Q are predicates recursive in B. If C is Δ_2^0 in B then $C \leq_T B'$. The latter is Post's Theorem. (See Rogers (7) pg. 303-4, and Theorem VIII, pg.314.)

One way to do this is to embed C into B on the primes in a Δ_2^0 way. We will embed $C(m)$ into B via

$m \in C \rightarrow$ a final segment of $\{p_m^k\}_{k>0}$ is in B

$m \notin C \rightarrow$ a final segment of $\{p_m^k\}_{k>0}$ is out of B.

We'll show that this is Δ_2^0 in B (pg. 72). Then $\underline{c} \leq \underline{b}'$.

From Chapter III we have a method which we can use to push the jump of B down: If we can construct B by stages β_s where $\{\beta_s\}_{s>0}$ is recursive in $\underline{a}$ then $B = \lim_s \beta_s \leq_T C$ (this is Shoenfield's Lemma relativized to $\underline{a}$). If we can make the jump of B settle down as we did in Chapter III, that is $\lim_s J_s(\beta_s, e) = J(B,e)$, $\forall e$, then we will have $B' \leq_T C$, again by the Limit Lemma, as $\{\beta_s\}_{s>0}$ is recursive in $\underline{a}$.

We need the full approximation construction because we will want $B = \lim_s \beta_s$, $\{\beta_s\}_{s>0}$ recursive in $\underline{a}$. Yet we can't just take the construction of Chapter III and relativize it to $\underline{a}$ (i.e. push everything above $\underline{a}$). For then we will get T_e^* partial recursive in $\underline{a}$, from which we can only conclude that $\underline{B}$ is a minimal cover for $\underline{a}$ ($\not\exists$ $\underline{d}$, $\underline{a} < \underline{d} < \underline{b}$), rather than that $\underline{B}$ is minimal.

What we will do, roughly speaking, is construct $\{\beta_s^\sigma\}$ at stage s for each string σ. This we can do uniformly recursively. We will try to make $J_s(\beta_s^\sigma, lh(\sigma))$ settle down, and to embed σ along the primes into all sufficiently long extensions of β_s^σ. Then when we're all done we'll take $B = \lim_s \beta_s^{\sigma_s}$ where σ_s is an $\underline{a}$ recursive approximation to C. To do this we need a tree of trees.

A tree of trees is nothing more than a tree, each of whose nodes is itself a tree. That is, for all strings σ, we will construct T_σ.

T_σ we will construct by stages $T_{\sigma,s}$. $\{T_{\sigma,s}\}_{s\geq 0}^{\sigma \text{ a string}}$ will be recursive. $T_{\sigma,s}$ will be related to our old trees $T_{e,s}$, via $lh(\sigma) = e$, then $T_{\sigma,s}$ will be an approximation to an e-splitting tree (think of $T_{\sigma,s}$ as $T_{e,s}^\sigma$). The compatibility requirements will now be: $T_{\tau,s}(\emptyset) \not\supseteq T_{\sigma,s}(\emptyset)$ if both are defined and $\tau \supset \sigma$. And $T_{\tau,s}$ will be compatible with $T_{\sigma,s}$. $T_{\emptyset,s}$ will play the role of our old $T_{0,s}$: it will be the full recursive tree from which all choices of strings will be made at s.

IV-1

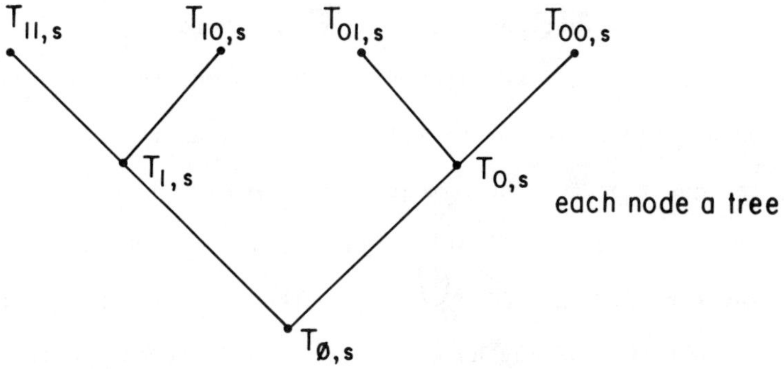

Essentially $T_{\sigma*i,s}(\emptyset) = T_{\sigma,s}(i)$ $i \in \{0,1\}$. So the nesting of trees will look like:

IV-2 (a partial diagram only)

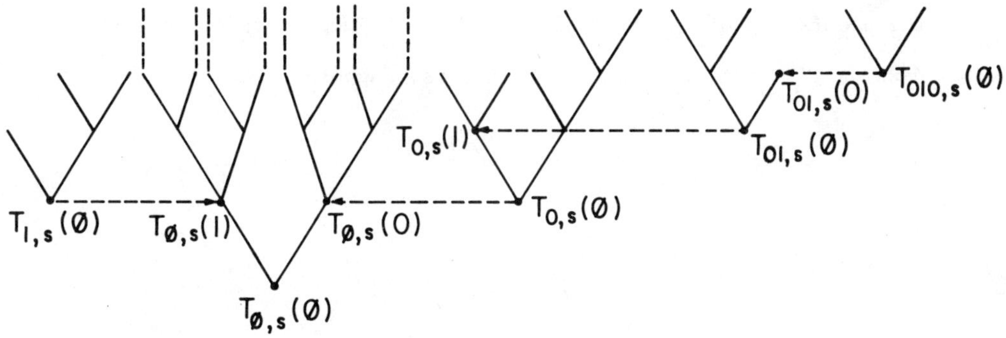

Explicitly, once we have $T_{\sigma,s}(\emptyset)\downarrow$, then if $lh(\sigma) = e$, $T_{\sigma,s}(\tau)$ is defined exactly as $T_{e,s}(\tau)$ in $\underline{m}' = \underline{0}'$ replacing e with σ everywhere, except that where we refer to compatibility with, or boundary strings for $e' < e$ there, we replace it with compatibility with, or boundary strings for $T_{\tau,s}$ $lh(\tau) = e' < e$, and $\underline{\underline{\tau \subset \sigma}}$. Prohibitions will be just as there.

If we can make $T_{\sigma,s}(\emptyset)$ settle down, then $T_{\sigma,s}(\tau)$ will settle down $\forall \sigma, \tau$. The proof will be exactly as in $\underline{m}' = \underline{0}'$ (we're begging the question for the moment of how we define $T_{\sigma,s}(\emptyset)$, but it will be, except for diagonalization, just as in $\underline{m}' = \underline{0}'$). Then if we take <u>any</u> set D, and $G = \bigcup_{\sigma \subset D} T_\sigma(\emptyset)$ then $\underline{G}$ will be a minimal degree (except that we may have $\underline{G} = \underline{0}$). Why? We can define T_e^* as in $\underline{m}' = \underline{0}'$ by everywhere replacing $T_{m,s}$ $m \leqslant e$ in that definition with $T_{\tau,s}$, $\tau \subset D$, $lh(\tau) = m$. B lies along the beginnings of these trees. They are compatible and $T_{\tau,s}$ is an m-splitting tree approximation. Thus all the salient features of Lemma 3 carry over and we can prove that T_e^* has properties (1)-(6).

So how do we define $T_{\sigma,s}(\emptyset)$, and why do we keep referring to the proofs of $\underline{m}' = \underline{0}'$? Well, we want the jump operator to settle down on $T_{\sigma,s}(\emptyset)$ for each σ. That is, suppose we have:

<u>IV-3</u>

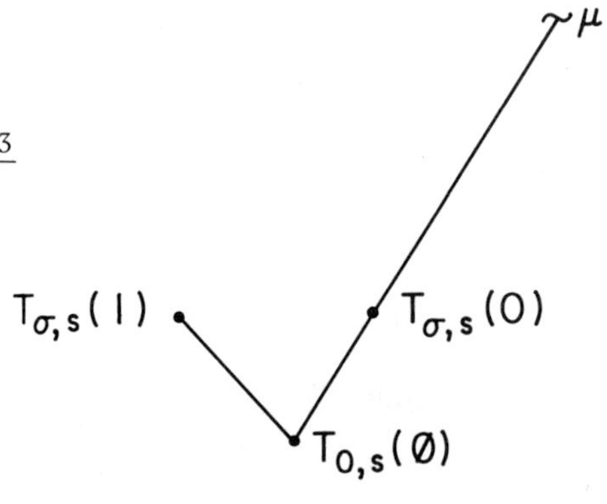

and we find $\mu \supset T_{\sigma,s}(0)$, compatible with $T_{\tau,s}$ $\forall \tau < \sigma$ and $J_s(\mu, lh(\sigma * 0) > 0$. Then we will make μ a follower of $T_{\sigma,s}(0)$ as in $\underline{m}' = \underline{0}'$ (recall that we don't do any diagonalization here).

Then, again as in $\underline{m}' = \underline{0}'$, we place $T_{\sigma * 0,t}(\emptyset)$, $t \geq s$ above μ, by making $T_{\sigma * 0,t}(\emptyset)$ = least string $\supseteq \mu$ compatible with $T_{\tau,t}\ \forall \tau \subseteq \sigma$.

IV-4

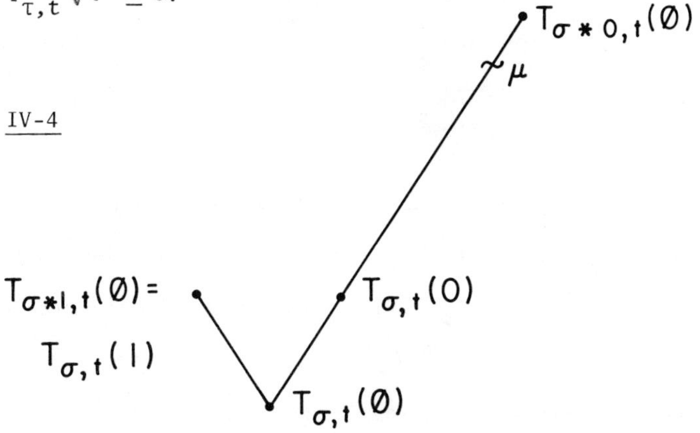

If we can ever do the same for $T_{\sigma,t}(1)$ and $T_{\sigma*1,t}(\emptyset)$ we will. Otherwise $T_{\sigma*1,t}(\emptyset) = T_{\sigma,t}(1)$. Thus just as in $\underline{m}' = \underline{0}'$, limits will exist. All we've done is expand the $\underline{m}' = \underline{0}'$ construction. What we get for this is:

if $B \supset T_\sigma(\emptyset)$, $lh(\sigma) = e$

then $\lim_s J_s(T_{\sigma,s}(\emptyset),e) = J(T_\sigma(\emptyset),e)$

$= J(B,e)$,

as we did in $\underline{m}' = \underline{0}'$.

If we take $\sigma_s = \max \sigma$ such that $\sigma \subset C^s$ and $T_{\sigma_s,s}(\emptyset)\downarrow$, then let

$$\beta_s = T_{\sigma_s,s}(\emptyset), \quad B = \lim \beta_s,$$

and see what we have. This will be equivalent to $B = \bigcup_{\sigma \subset C} T_\sigma(\emptyset)$, since $\sigma \subset C \to \sigma \subset C^s$ all sufficiently large s. So

$\lim_s J_s(\beta_s,e) = \lim_s J_s(T_{\sigma_s,s}(\emptyset),e)$

(rec. ↑ J_s; rec. in $\underline{a}$ ↑ β_s)

$= \lim_s J_s(T_{\sigma,s}(\emptyset),e)\quad lh(\sigma) = e,\ \sigma \subset C$

$= J(B,e),\ \forall e.$

Hence $J(B,e)$ is recursive in $C \in \underline{a}'$. That is $B' \leq_T C$.

Note that σ_s and β_s are defined externally to the construction of our tree of trees. σ_s and β_s do not affect the construction in any way. They just allow us to pick a path through the trees. This is essential for us to be able to obtain $T_{\sigma,s}$ recursive $\forall \sigma, s$.

We are left now with pushing the jump of B up. To do this we will play with $T_{\emptyset,s}$. Since all we need for our above description to work is $T_{\emptyset,s} \subseteq T_{\emptyset,s+1}$ and $T_\emptyset$ full, recursive, we have some leeway in defining $T_{\emptyset,s}$.

First we need to make a definition. The rank of a string $\underline{\delta}$ at stage $\underline{s}$ is:

$$R(\delta,s) = \sigma \text{ where } \sigma \text{ is the longest string such that}$$

$$T_{\sigma,s}(\emptyset)\downarrow \subseteq \delta.$$

How do we define $T_{\emptyset,s}$ to effect a primes embedding?

If $T_{\emptyset,s-1}(\tau)\downarrow$ set $T_{\emptyset,s}(\tau) = T_{\emptyset,s-1}(\tau)$ and define $T_{\emptyset,s}(\tau * 0)(\tau * 1)$.

To define $T_{\emptyset,s}(\tau * 0)(\tau * 1)$ suppose $R(T_{\emptyset,s}(\tau)) = \sigma$. Let σ_0,σ_1 be the least strings such that $\sigma_0,\sigma_1 \supseteq T_{\emptyset,s}(\tau)$ and

$$\sigma_0|\sigma_1,\ lh(\sigma_0) = lh(\sigma_1),$$

and if $\sigma(m)\downarrow$ then

$$lh(T_{\emptyset,s}(\tau)) \leq p_m^k < lh(\sigma_i)\ i = 0,1$$

$$\Longrightarrow \sigma_0(p_m^k) = \sigma_1(p_m^k) = \sigma(m).$$

Set $T_{\emptyset,s}(\tau * 0)(\tau * 1) = \sigma_0,\sigma_1$.

IV-5

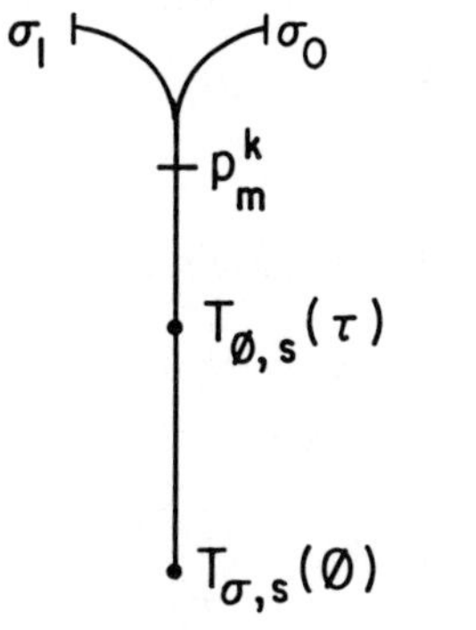

$\sigma(m) = \sigma_0(p_m^k) = \sigma_1(p_m^k)$

This will give us $T_\emptyset$ recursive.

What we have done is embed $\sigma(m)$ onto all sufficiently large p_m^k in all sufficiently long extensions of $T_\sigma(\emptyset)$, for every σ. That is, if $B \supset T_\sigma(\emptyset)$ then for every $m < lh(\sigma)$

if $\sigma(m) = 0$, a final segment of $\{p_m^k\}_{k>0}$ is in B,

if $\sigma(m) = 1$, a final segment of $\{p_m^k\}_{k>0}$ is out of B.

To see this just let s_0 be such that $\forall s > s_0 \; T_{\sigma,s}(\emptyset) = T_{\sigma,s_0}(\emptyset)$. Let $T_{\emptyset,s_0}(\pi)$ be an end string on $T_{\emptyset,s_0}$ such that $T_{\emptyset,s_0}(\pi) \subset B$. Then $\forall s > s_0$, $R(T_{\emptyset,s}(\pi * \lambda), s) \supseteq \sigma$ for every $\lambda \supseteq \emptyset$.

IV-6

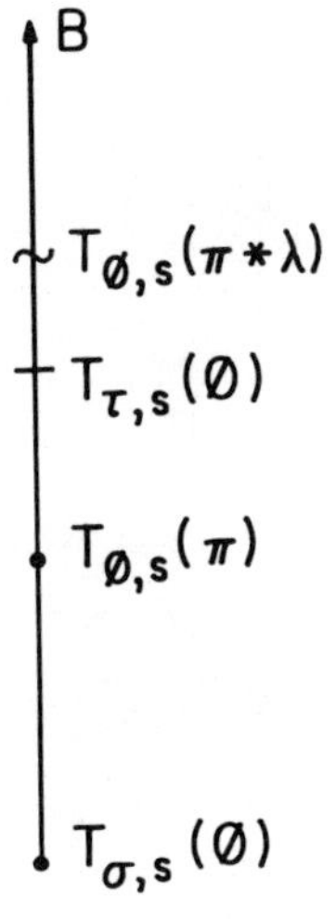

$\sigma \subseteq B$

So when we define $T_{\emptyset,s}(\pi * \lambda)(p_m^k)$ for any k such that $p_m^k > lh(T_{\emptyset,s}(\pi))$, $T_{\emptyset,s}(\pi * \lambda)(p_m^k) = \tau(m) = \sigma(m)$ so $\forall k$ such that $p_m^k > lh(T_{\emptyset,s}(\pi))$, $B(p_m^k) = \sigma(m)$.

Since we will have $B = \bigcup_{\sigma \subset C} T_\sigma(\emptyset)$, we will have

$$m \in C \leftrightarrow \exists x \ \forall y > x (p_m^y \in B)$$

and also

$$m \in C \leftrightarrow \forall x \ \exists y > x (p_m^y \in B)$$

the latter because either all p_m^y's or no $p_m^y \in B$ for sufficiently large y.

Hence C is Δ_2^0 in B. Thus $C \leqslant_T B'$. So together with our earlier discussion, for the B we so construct

$$B' \equiv_T C,$$

and $\underline{B}$ is minimal.

This concludes the description.

Further Topics and Remarks

Remember that we construct our tree of trees without reference to β_s. Hence for every $\underline{c} > \underline{0}'$, $C \in \underline{c}$ we may choose a

$$B = \bigcup_{\sigma \subset C} T_\sigma(\emptyset)$$

from our tree of trees such that

$\underline{b}$ is minimal and $\underline{b}' = \underline{c}$.

Thus all the minimal degrees necessary to prove our theorem are exhibited at once.

Another feature of this construction is that

$$\underline{b}' = \underline{b} \cup \underline{0}'$$

Why? For any σ, recursive in $0'$ we can find $T_\sigma(\emptyset)$. (Look for $\mu_s(\forall t > s\ T_{\sigma,t}(\emptyset) = T_{\sigma,s}(\emptyset))$. Then to compute $C(x)$, $C \in \underline{c} = \underline{b}'$ look for σ such that $lh(\sigma) > x$ and $T_\sigma(\emptyset) \subset B$. Then $C(x) = \sigma(x)$. Thus not only do we have a minimal $\underline{b}$ such that $\underline{b}' = \underline{c}$, but $\underline{b}$ has least possible jump. This was first noted by L. Sasso.

Is it possible that for every minimal $\underline{m}$, $\underline{m} \cup \underline{0}' = \underline{m}'$? No, for L. Sasso has recently exhibited in (9) a minimal degree $\underline{m}$ such that $\underline{m} \cup \underline{0}' \neq \underline{m}'$. This does not require a priority argument. We recommend (9) to the reader as a very clear proof.

Even after that was settled an earlier question of Yates (13, p. 265) as to whether there was a minimal $\underline{m} < \underline{0}'$ such that $\underline{m}' > \underline{0}'$ remained. Sasso, Cooper and we have recently observed how to modify the construction of (9) to show that such an $\underline{m}$ exists. This requires only an e-state construction, and is outlined in (9).

It is natural then to conjecture that for every $\underline{c}$ r.e. in $\underline{0}'$

there is some $\underline{m}$ minimal, $\underline{m} < \underline{0}'$ such that $\underline{m}' = \underline{c}$. However, Cooper (2 , Theorem 2) has shown that if $\underline{d} \leqslant \underline{0}'$ and $\underline{d}' = \underline{0}''$ ($\underline{d}$ is "high") then $\underline{d}$ is not minimal. In (2 , Theorem 3) he improves this to show that if $\underline{d}$ is high, there is some $\underline{m}$ minimal, $\underline{m} < \underline{d}$. This requires the full approximation construction.

It is an open question now what, if any, classification there is for the jumps of minimal degrees less than $\underline{0}'$. Jockusch, we believe, has suggested that $\underline{m} < \underline{0}'$, $\underline{m}$ minimal $\rightarrow \underline{m}'' = \underline{0}''$. We believe that if it is possible to construct a minimal $\underline{m} < \underline{0}'$ with prescribed jump $> \underline{0}'$ it will require a full approximation to that $\underline{m}$.

CHAPTER V: A MINIMAL DEGREE $\underline{m} < \underline{a}$ r.e.

Sections

* * *

In this Chapter we prove that given $\underline{0} < \underline{a}$ recursively enumerable (r.e.), then there is a minimal degree $\underline{m} < \underline{a}$. As we noted at the beginning of Chapter III, this theorem is due to Yates (13). Our proof is significantly different because we base it on the full approximation construction of Chapter II.

We believe that to the reader who has a reasonable understanding of Chapter II this construction will be very intuitive. For that reason we give an extensive motivation. It is our hope that this will be sufficiently clear to enable the reader to see how to construct such a minimal degree without actually going through our construction and proof in detail. The only ideas found in the construction and proof which are not explained in the motivation concern the order of proof. This we set out as the "plan of proof."

We shall always assume $A \in \underline{a}$ r.e., $\underline{a} > \underline{0}$ and $f(\mathbb{N}) = A$, f 1-1 recursive.

It should be noted that no r.e. degree is minimal. Hence if $\underline{m}$ is minimal, $\underline{m} \leqslant \underline{a}$ r.e., then $\underline{m} \neq \underline{a}$. (see e.g. [11] §14).

Permitting

We first give an explanation of a simple technique at the heart of our construction: permitting.

Permitting is a technique which enables us to construct a degree, not necessarily r.e., less than a given non-recursive r.e. degree.

Suppose we are given A and f as before. How do we assure that in whatever construction we are doing the set B which we construct will satisfy $B \leqslant_T A$?

We will construct B by recursive stages, $\lim_s \beta_s = B$. Then we'll require that if

$$\forall s > t,\ f(s) \geqslant x \quad \text{then} \quad B[x] = \beta_t[x].$$

That is, once f no longer enumerates elements less than x we will fix B to level x. To compute B(x) we'll simply find a t as above and see what $\beta_t(x)$ is. Simple, no?

Of course balanced against this are whatever other conditions we wish B to satisfy. The best example of how we cope with them is the first part of our motivation concerning T_1.

In Chapter VI we will see how to use a variant of permitting to construct a degree $< \underline{0}'$ in a given relationship (other than $\leqslant$) to a given r.e. degree.

Motivation

To construct B of minimal degree, $B \leqslant_T A$, we will attach a permitting argument to our full approximation construction.

As usual we'll construct β_s as the longest beginning of any tree, and $\lim_s \beta_s = B$. Only now we'll allow $\beta_{s+1}(x) \neq \beta_s(x)$ only if $f(s+1) < x$. How are we to do this and still make B minimal?

First let's give a name to the string we must preserve as a beginning of B at stage s+1:

> The retraced path at s+1 δ_{s+1} is the maximal string $\delta_{s+1} \subseteq \beta_t$ some $t \leqslant s$ such that there has been no application of permitting at $d < lh(\delta_{s+1})$ at any stage r, $t < r \leqslant s$.

An application of permitting will entail $f(r) = d$. Hence δ_{s+1} will be the longest string which we must retain as $\subseteq \beta_{s+1}$. The reason we do not require $\delta_{s+1} = \beta_t$ some $t \leqslant s$ will become clearer as we go along.

In our full approximation construction how does β_{s+1} become incompatible with β_s? The first and most important way is by defining new Case II (splitting) extensions on $T_{e,s}$.

V-1

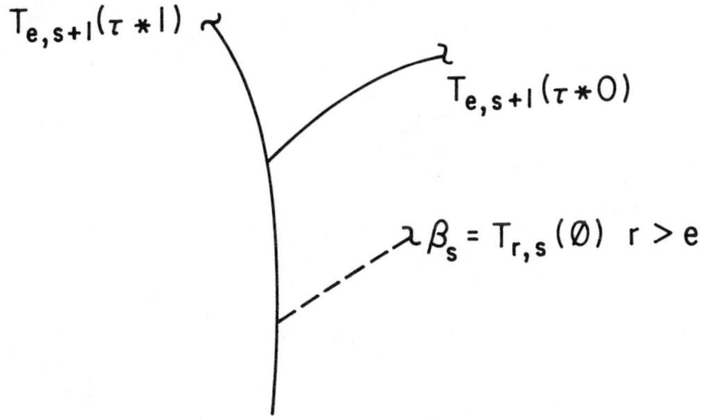

We must not allow this to happen unless f permits it. Looking first at T_1, consider $T_{1,s}(\tau)$ an end string on $T_{1,s}$. Recall that $T_{1,s}(\tau) = T_1(\tau)$ always. We will try to find extensions for $T_1(\tau)$ at stage s only if $T_{1,s}(\tau) \subseteq \beta_{s-1}$ (see page 20). This will considerably simplify our arguments. Now suppose we find a 1-splitting (σ_1,σ_2) above $T_1(\tau)$ and $T_1(\tau) \subseteq \beta_{s-1}$.

V-2

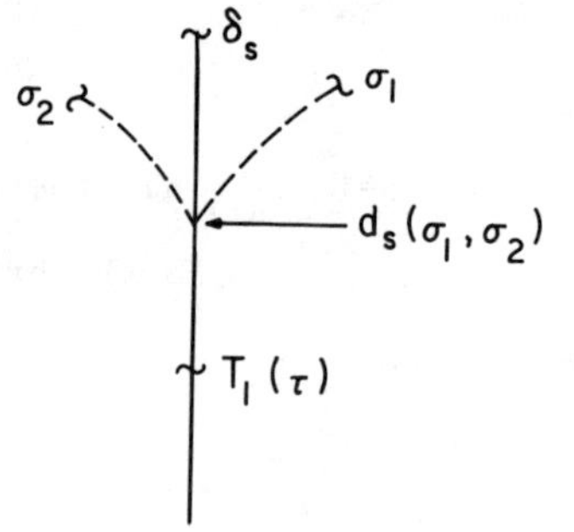

Let $d_s(\sigma_1,\sigma_2) = \mu y([\sigma_1(y) \neq \sigma_2(y)] \vee [\sigma_1(y) \neq \delta_s(y)] \vee [\sigma_2(y) \neq \delta_s(y)])$.

We call $d_s(\sigma_1,\sigma_2)$ the <u>divergence point of σ_1,σ_2 from the retraced path at s</u>. If we ever use σ_1,σ_2 as a branching then we will say that we have <u>created a divergence point at</u> $d_s(\sigma_1,\sigma_2)$.

We have spotted a 1-splitting (σ_1,σ_2) which we would like to use. Erect a marker $(1, d_s(\sigma_1,\sigma_2), s)$ at $r = \delta_s(d_s(\sigma_1,\sigma_2))$.

V-3

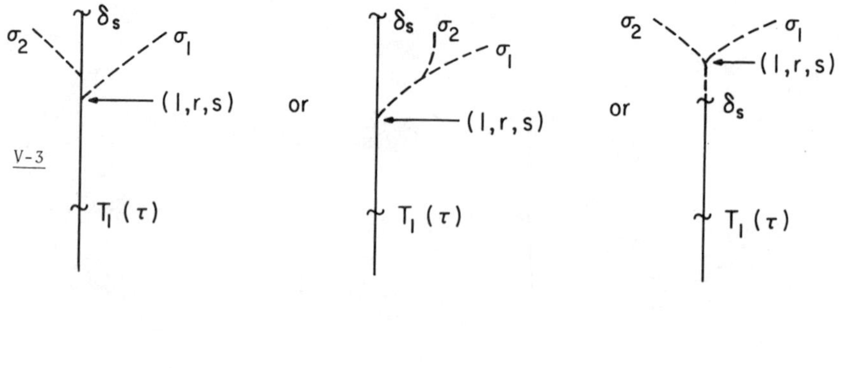

We'll keep (σ_1,σ_2) as a potential extension of $T_1(\tau)$ so long as $\beta_u \supseteq T_1(\tau)$, $u > s$. We'll use $\sigma_1,\sigma_2 = T_{1,u}(\tau * 0)(\tau * 1)$ only if we have $f(u) < d_s(\sigma_1,\sigma_2)$, that is, if f permits us to. For then it is all right to change β_{u+1} above r. But we may never have $f(u) < r$. So we'll look for another 1-splitting τ_1,τ_2 (so call $(1,r,s) = (1,r_1,s_1)$). If we find one at stage s_2, $\beta_{s_2-1} \supseteq T_1(\tau)$,

V-4

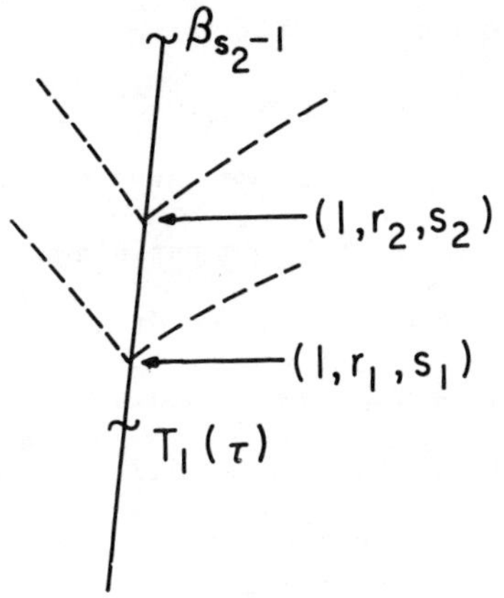

then we'll attach a marker $(1,r_2,s_2)$ to it, r_2 defined as r_1 was. We'll require $r_2 > r_1$. If $r_2 \leqslant r_1$ this new splitting would be of no more use, relative to f, than the first one.

We continue in this manner finding potential extensions of $T_1(\tau)$.

V-5

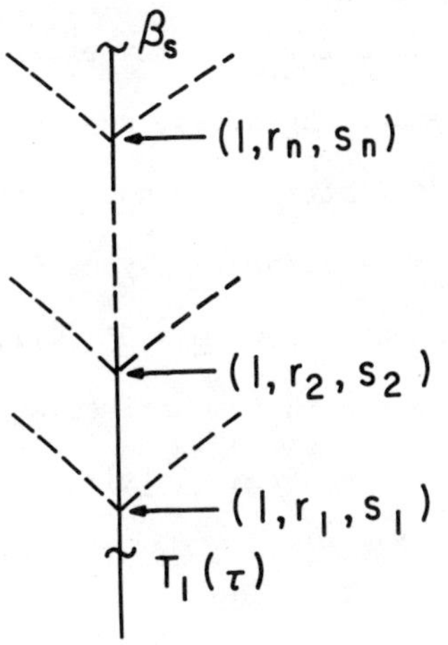

If $B \supset T_1(\tau)$ and there are 1-splittings lying above arbitrarily long beginnings of B then <u>we will be permitted to use one of them as $T_1(\tau * 0)(\tau * 1)$</u>. If we couldn't we'd get an infinite sequence of markers $(1,r_n,s_n)_{n=1}^{\infty}$ with $r_n < r_{n+1}$, $s_n < s_{n+1}$, which would be recursive. Never being permitted to use any splitting we'd have $\forall n$, $\forall s > s_n$, $f(s) > r_n$. But this would imply that A is recursive: to compute $A(x)$ find an n such that $r_n > x$; then $x \in A \leftrightarrow x = f(s)$ some $s \leqslant s_n$. That is a contradiction.

What about T_e? There we have to consider that potential extensions may no longer be compatible with $T_{r,s}$ $\forall r < e$. How do we pull out a recursive sequence of markers as above? We proceed just as for T_1, except we retain a potential extension only if it continues to satisfy the usual conditions of Case II (compatible, no intervening boundary strings) and $\beta_{u-1} \supseteq T_{e,u}(\tau) = T_{e,s}(\tau)$, s the stage at which appointed. The potential extensions which will be permanent will be those that lie on T^*_{e-1}. T^*_{e-1} is partial recursive, so we will be able to find the permanent potential extensions recursively.

What happens if boundary strings intervene between $T_{e,s}(\tau)$ and a potential extension? We go ahead and define $T_{e,s}(\tau * 0)(\tau * 1)$ as a dummy extension as before. Dummy extensions will never alter the path of β_s; they are harmless, and their only purpose is to lift T_e above boundary strings. Thus Cases I and III will remain unchanged. Case II is as before, only used less frequently. Therefore many of our arguments from $\underline{m} < \underline{0}'$ will carry over intact.

The reader may notice that sometimes even though we do not create splittings diverging from $\beta_s[x]$ if $f(s+1) > x$, we may still find that $\beta_{s+1}(x)\not\downarrow$.

V-6

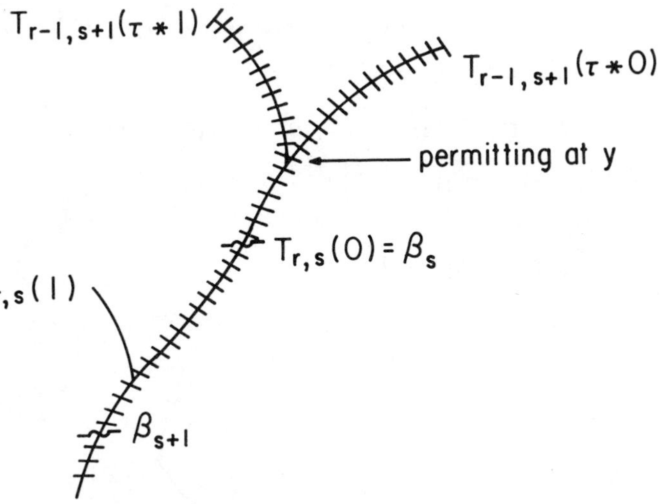

This will be no problem since both possible paths above β_{s+1} are compatible with $\beta_s[x]$, and must remain so due to our definition of δ_{s+1}.

Note that no cancellations will occur on any tree unless we have an application of permitting.

Lastly, what about making B not recursive? In $\underline{m} < \underline{0}'$ we switched $T_{e+1,s}(\emptyset) = T_{e,s}(1)$ if $T_{e,s}(0) \subseteq \phi_{e+1,s}$. Here we can do this only if we are permitted to by f. As in our arguments for splittings, we will need infinitely many chances to make $B \neq \phi_{e+1}$ to be sure of being able to use one. So we introduce new trees, $T^{rec}_{e,s}$, for the purpose of marking out these chances. We think of $T^{rec}_{e,s}$ as replacing $T_{e,s}(0)(1)$ in the diagonalization process.

We'll start $T^{rec}_{e,s}(\emptyset) = T_{e,s}(\emptyset)$. We'll define $T^{rec}_{e,s}(0^m * 0)(0^m * 1)$ as a least dummy extension of $T^{rec}_{e,s}(0^m)$. Here

$0^m = \underbrace{0 * 0 * \ldots * 0}_{\text{m-times}}$, $0^o = \emptyset$. A dummy extension of $T^{rec}_{e,s}(0^m)$ will just be a pair of strings extending $T^{rec}_{e,s}(0^m)$ compatible with $T_{r,s}$ $\forall r \leqslant e$.

We will define $T^{rec}_{e,s}(0^m * 0)(0^m * 1)$ only if $\beta_{s-1} \supseteq T^{rec}_{e,s}(0^m)$, making sure that $T^{rec}_{e,s}(0^m * 0)$ lies on the retraced path at s.

V-7

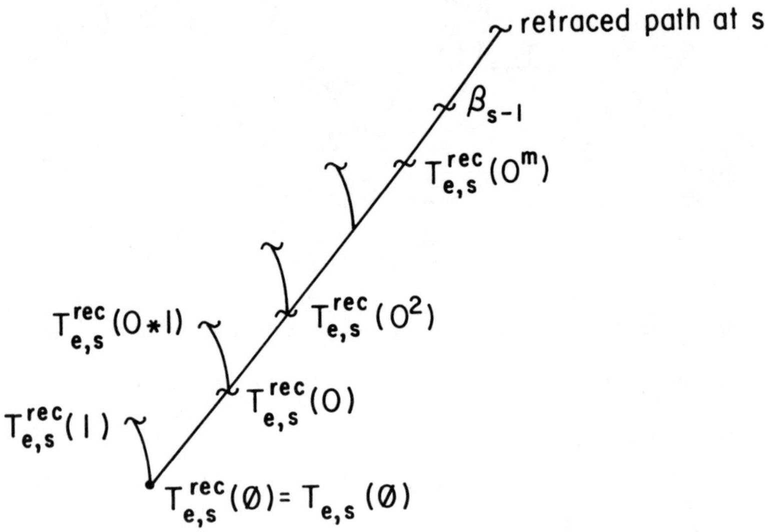

If $T^{rec}_{e,s}(i) \nsubseteq \phi_{e+1,s}$, $i \in \{0,1\}$, then we'll give $T^{rec}_{e,s}(0)(1)$ a marker (e rec, $lh(T^{rec}_{e,s}(\emptyset))$,s). Likewise, if all the previous branchings have markers, and $T^{rec}_{e,s}(0^m * 1) \nsubseteq \phi_{e+1,s}$ $i \in \{0,1\}$, we'll give $T^{rec}_{e,s}(0^m * 0)(0^m * 1)$ a marker (e rec, $lh(T^{rec}_{e,s}(0^m))$,s).

V-8

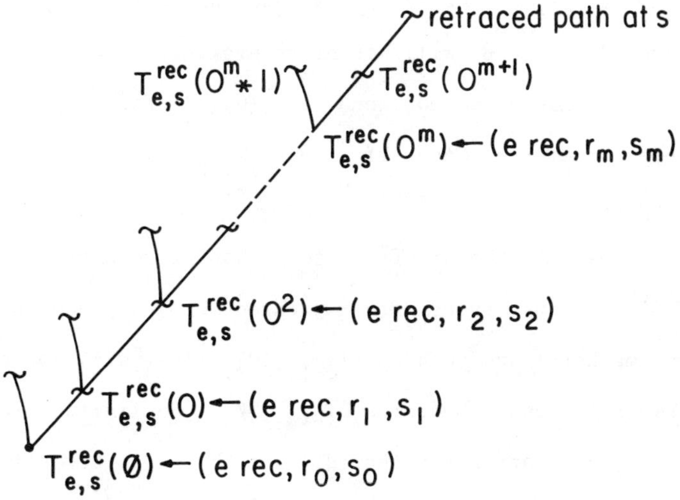

If $f(s) < r_m$, m minimal, then we'll use $T^{rec}_{e,s}(0^m * 0)(0^m * 1)$ to make $B \neq \phi_{e+1}$, by placing $T_{e+1,s}(\emptyset) = T^{rec}_{e,s}(0^m * i) \not\subseteq \phi_{e+1,s}$. f permits us to change β_s above $T^{rec}_{e,s}(0^m)$. The marker $(e\ rec, r_m, s_m)$ is then valid for e.

If we have several markers up but none valid, we'll set $T_{e+1,s}(\emptyset) = T^{rec}_{e,s}(0^m * 0)$, m maximal such that $T^{rec}_{e,s}(0^m * 0)(0^m * 1)$ has a marker. That will push β_s above $T^{rec}_{e,s}(0^m * 0)$ and keep it there, so that we can define $T^{rec}_{e,t}(0^{m+1} * 0)(0^{m+1} * 1)$, $t > s$. It will also keep β_s on the retraced path at s.

If ϕ_{e+1} is total we'll eventually be able to use one of these branchings to make $B \neq \phi_{e+1}$. If we couldn't we'd get an infinite recursive sequence of markers as we did for splittings above, leading to the contradiction that A is recursive. To show that the sequence is recursive will require a little work. If we couldn't use one of the branchings, $T_{e+1,s}(\emptyset)$ would continue to be pushed up so that we would be able to define $B = \bigcup_{m \geq 0} T^{rec}_e(0^m)$. Then we could use this B to define T^*_e. We can then locate the permanent markers of

T_e^{rec} on T_e^*. (The reader can see this more fully in the plan of proof, page 89). So B will not be recursive.

Because we do not need $T_{e,s}(0)(1)$ for diagonalization we will delete them from Case III. Thus $T_{e,s}$ will be concerned only with e-splitting.

How will the proofs of this Chapter compare to $\underline{m} < \underline{0}'$? The $T_{e,s}^{rec}$ are just trees of dummy extensions, a place on which to hang our markers, and push up $T_{e+1,s}(\emptyset)$. They will not affect $T_{r,s}$ any r, except to define $T_{e+1,s}(\emptyset)$. So limits, except for $T_{e+1,s}(\emptyset)$, are not affected. The T_e^* will be defined exactly as before. Properties (1)-(6) of T_e^* will be proved exactly as before except for (2) (page 35). (2) requires that we can always put up e-splittings if necessary; that we have already shown. So we will have that B is minimal. That $B \leqslant_T A$ is the whole point of our permitting argument, and will follow easily.

Well, that ends the motivation. You're now permitted to go on. (Permitted? Who, me?).

Construction of $\underline{m} < \underline{a}$ r.e.

Let $A \in \underline{a} > \underline{0}$, f recursive, 1-1, $f(\mathbb{N}) = A$.

$T_{0,s}$ = identity tree to level s.

As usual, all strings are chosen from $T_{0,s}$ at stage s.

$T_{e,s}$, $e > 0$ $T_{e,s}(\emptyset)$: If $T_{e,s}(\emptyset)$ has been placed at $T^{rec}_{e-1,s}(\gamma)$, then set $T_{e,s}(\emptyset) = T^{rec}_{e-1,s}(\gamma)$.

Otherwise let m be maximal such that $T^{rec}_{e-1,s}(0^m * 0)(0^m * 1)$ has a marker.

Set $T_{e,s}(\emptyset) = T^{rec}_{e,s}(0^m * 0)$.

If there is no such m, stop.

We continue only if $T_{e,s}(\tau)\downarrow = T_{e,s-1}(\tau)$.

Case I: $T_{e,s-1}(\tau * 0)(\tau * 1)\downarrow$ and compatible with $T_{r,s}$ $\forall r < e$,

and [(i) $\tau = \emptyset$] ← delete

or (ii) $T_{e,s-1}(\tau * 0)(\tau * 1)\downarrow$ originally by Case II and (b) of Case II potential still holds.

or (iii) $\nexists \sigma_1, \sigma_2$ as in Case II; or $\beta_{s-1} \not\supseteq T_{e,s}(\tau)$.

Then set $T_{e,s}(\tau * 0)(\tau * 1) = T_{e,s-1}(\tau * 0)(\tau * 1)$.

We continue only if $T_{e,s}(\tau) \subseteq \beta_{s-1}$.

Case II (potential): If $T_{e,s}(\tau * 0)(\tau * 1)\downarrow$ it is not originally by Case II, and $\exists \sigma_1, \sigma_2$ such that

(a) σ_1, σ_2 are compatible with $T_{r,s}$ $\forall r < e$.

(b) Every $\sigma' \subset \sigma_1$ or $\sigma' \subset \sigma_2$ which is a boundary string for some $r < e$ satisfies $\sigma' \subsetneq T_{e,s}(\tau)$.

(c) σ_1, σ_2 e-split $T_{e,s}(\tau)$ at s.

(continued)

(d) If $r = d_s(\sigma_1,\sigma_2)$, and the marker for the last appointed current potential extension of $T_{e,s}(\tau)$ is (e,r',t), then $r' < r$ (or there is no such marker).

Then we appoint the least such (σ_1,σ_2) a potential extension of $T_{e,s}(\tau)$ and associate with it the marker $(e,d_s(\sigma_1,\sigma_2),s)$.

It is a current potential extension of $T_{e,u}(\tau)$, $u \geqslant s$ only if $T_{e,u}(\tau) = T_{e,u-1}(\tau)$ and (a), (b), (c) of Case II (potential) still hold for (σ_1,σ_2) with s replaced by u, and $\beta_{u-1} \supseteq T_{e,u}(\tau)$. It is cancelled if any of these conditions fail, or if $T_{e,u}(\tau * 0)(\tau * 1)\downarrow$ by Case II.

We use Case II (potential) at most once at stage s for $T_{e,s}(\tau)$.

Note that by itself it does not define an extension at stage s.

Case II: Case I does not apply and $\exists$ a current potential extension (σ_1,σ_2) of $T_{e,s}(\tau)$ with marker (e,r,t) and $f(s) < r$.

Then set $T_{e,s}(\tau * 0)(\tau * 1)$ = the least such σ_1,σ_2. This is an application of permitting at r.

Case III: Cases I, II and II (potential) do not hold, and

either [(i) $\tau = \emptyset$)] ← delete

or (ii) $\exists \sigma_1,\sigma_2$ as in Case II (potential) except that (b) fails.

Then set $T_{e,s}(\tau * 0)(\tau * 1)$ = least dummy extension of $T_{e,s}(\tau)$.

$\underline{T^{rec}_{e,s}}$, $e > 0$: $T^{rec}_{e,s}(\emptyset) = T_{e,s}(\emptyset)$.

We continue only if $T^{rec}_{e,s}(\tau) = T^{rec}_{e,s-1}(\tau) \subseteq \beta_{s-1}$ and $\tau = 0^m$ some m.

(A) No branching $T^{rec}_{e,s}(\gamma * 0)(\gamma * 1)$ has a valid marker for e, $lh(\gamma) < lh(\tau)$.

Let σ_0, σ_1 = least dummy extension of $T^{rec}_{e,s}(0^m)$.

Set $T^{rec}_{e,s}(0^m * 0)(0^m * 1) = \sigma_i, \sigma_{1-i}$ where i is minimal such that σ_i lies on the retraced path at s.

We say that we have "chosen $T^{rec}_{e,s}(0^m * 0)(0^m * 1)$ by the retracing procedure."

We now continue with Case (B) or (C) for $T^{rec}_{e,s}(0^m * 0)(0^m * 1)$ if we can. Neither define extensions on $T^{rec}_{e,s}$.

(B) $T^{rec}_{e,s}(0^m * 0)(0^m * 1)$ has a marker (e rec, r, t) and $f(s) < r$.

Then <u>place</u> $T_{e+1,s}(\emptyset)$ at $T^{rec}_{e,s}(0^m * i)$ where i is minimal such that $T^{rec}_{e,s}(0^m * i) \not\subseteq \phi_{e+1,s}$.

$T_{e+1,u}(\emptyset)$ continues to be placed at $T^{rec}_{e,u}(0^m * i)$, $u > s$, if $T^{rec}_{e,u}(0^m * 0)(0^m * 1) = T^{rec}_{e,u-1}(0^m * 0)(0^m * 1)$. Otherwise this placement is cancelled.

(e rec, r,t) is now a <u>valid</u> marker for e.

It stays so until the placement is cancelled.

This is an <u>application of permitting at r</u>.

(C) Every branching $T^{rec}_{e,s}(\gamma * 0)(\gamma * 1)$, $\gamma \subsetneq 0^m$ has a marker for e which is not valid, but $T^{rec}_{e,s}(0^m * 0)(0^m * 1)$ has no marker. And $T^{rec}_{e,s}(0^m * i) \not\subseteq \phi_{e+1,s}$ $i = 0$ or 1. Then set up a marker for this branching (e rec, $lh(T^{rec}_{e,s}(0^m))$,s). This marker remains up at stage $u > s$ if $T^{rec}_{e,u}(0^m * 0)(0^m * 1) =$

$T^{rec}_{e,u-1}(0^m * 0)(0^m * 1)$. Otherwise it is cancelled.

If Case (B) or (C) is used for $T^{rec}_{e,s}(0^m * 0)(0^m * 1)$ stop construction on $T^{rec}_{e,s}$ and proceed to $T_{e+1,s}$. Otherwise continue on $T^{rec}_{e,s}$.

* * *

At the end of stage s set

$$m_s = \max m \ (T_{m,s}(\emptyset)\downarrow)$$

and $\beta_s = T_{m_s,s}(\emptyset)$.

*

Plan of Proof

Since we want to invoke some of our proofs from $\underline{m} < \underline{0}'$ we must first point out the similarities to the reader.

The most important similarity is between the new and old $T_{e+1,s}$. Case I, II, and III are the same as before, except that we may invoke Case II less frequently, so that extensions on $T_{e+1,s}$ are defined as before. However, $T_{e+1,s}(\emptyset)$ is defined considerably differently. But once we show that $T_{e+1,s}(\emptyset)$ settles down we will be able to claim that $\forall \tau$, $\lim_s T_{e+1,s}(\tau)$ exists by invoking the proof of Lemma 1 of $\underline{m} < \underline{0}'$. This is because the $T^{rec}_{m,s}$, $m < e$ are just trees of dummy extensions. We will prove that $\lim_s T_{e+1,s}(\emptyset)$ exists by contradiction. We will assume that $\lim_s T_{m,s}$ and $\lim_s T^{rec}_{m,s}$ exist $\forall m \leqslant e$. If we assume that $T_{e+1,s}(\emptyset)$ does not settle down we will be able, nonetheless, to define $B = \lim_s \beta_s$. Using this B we will define T^*_m $\forall m \leqslant e$ as in Lemma 3 of $\underline{m} < \underline{0}'$ satisfying the same properties (1)-(6). Only (2) will require a new proof. We need T^*_e to recursively locate the permanent markers on T^{rec}_e. We will be able to define the T^*_m since they do not depend on T_r, $r > m$, for anything except that $B = \lim_s \beta_s$ exists. Recursively locating the markers will allow us to conclude that A is recursive, which will be the contradiction.

In outline, then, the plan of proof is:

Lemma A) (1) $\forall m \leq e$, $\forall \tau$, $\lim_s T_{m,s}(\tau)$ exists and $\lim_s T_{e+1,s}(\emptyset)\downarrow$

$\Rightarrow \forall \tau \lim_s T_{e+1,s}(\tau)$ exists.

— We will not expand on this any further.

(2) $\forall m \leq e$, $\forall \tau \lim_s T_{m,s}(\tau)$ exists

$\Rightarrow \forall \tau \lim_s T^{rec}_{e,s}(\tau)$ exists.

— We will prove this in detail.

Lemma B) $\forall m \leq e$, $\forall \tau \lim_s T_{m,s}(\tau)$ exists

$\Rightarrow \lim_s T_{e+1,s}(\emptyset)\downarrow$.

— All of this requires a detailed proof.

(i) If it were not we could still define $B = \lim_s \beta_s$.

(ii) Since we have B, we may define T^*_m, $\forall m \leq e$, as in Lemma 3 of $\underline{m} < \underline{0}'$. Properties (1)-(6) are immediate except for (2) which we prove.

(iii) We arrive at a contradiction using (i) and (ii).

—T^*_e allows us to recursively locate the permanent markers on T^{rec}_e.

Lemma C) B is not recursive.

— This will follow easily from Lemma B).

Lemma D) $\underline{b}$ is minimal.

— This follows immediately from Lemmas A) and B). From them we will have that $\lim_s T_{e,s}(\tau)$ exist, $\forall e$, so $\lim_s \beta_s = B$ exists. In Lemma B)(ii) we have already shown that we can construct the T^*_e appropriate to obtaining $\underline{b}$ minimal. Lemma D) will receive no further proof.

Lemma E) $B \leqslant_T A$

—We will give a computation procedure.

We proceed by proving the indicated parts of the lemmas.

Proof

Lemma A) (2) $\forall m \leqslant e$, $\forall \tau \lim_s T_{m,s}(\tau)$ exists
$\Rightarrow \forall \tau \lim_s T^{rec}_{e,s}$ exists.

Proof: We must show that if $\lim_s T^{rec}_{m,s}(\tau * 0)(\tau * 1)\downarrow$ infinitely often, then they converge. We induct on $lh(\tau)$.

$\lim_s T_{e,s}(\emptyset)\downarrow$. So $\lim_s T^{rec}_{e,s}(\emptyset)\downarrow = T_e(\emptyset)$.

Suppose $\lim_s T^{rec}_{e,s}(\tau)\downarrow$ and $T^{rec}_{e,s}(\tau * 0)(\tau * 1)\downarrow$ infinitely often. Then $\tau = 0^m$ some m.

Let s_0 be such that $\forall s > s_0$ $T^{rec}_{e,s}(0^m) = T^{rec}_{e,s_0}(0^m)$. Then some s_1, $\forall s \geqslant s_1$, $\beta_s \supseteq T^{rec}_{e,s}(0^m)$. Why? If not $T_{e+1,s}(\emptyset)$ is placed at $T^{rec}_{e,s}(0^{m-1} * 1)$. Once this happens, say at $s_2 > s_0$, it continues to occur at $\forall s > s_2$. But then we would not define $T^{rec}_{e,s}(0^m * 0)(0^m * 1)$ for infinitely many s.

Let $s_2 \geqslant s_1$ be such that the least dummy extension of $T^{rec}_{e,s}(0^m)$, $\forall s > s_2$, is well defined, say τ_0, τ_1. Some $\beta_s \supseteq T^{rec}_{e,s}(0^m)$, β_s is compatible with one of τ_0 or τ_1 $\forall s > s_2$. We can not apply permitting below $\ell = \max(lh(\tau_0), lh(\tau_1))$ infinitely often; say never after $s_3 \geqslant s_2$. So $\forall s > s_3$, β_s is compatible with τ_j, for one of $j = 0,1$, j minimal. So $\forall s > s_3$, $T^{rec}_{e,s}(0^m * 0)(0^m * 1) = \tau_j, \tau_{1-j}$. □

Lemma B) $\forall m \leqslant e$, $\forall \tau \lim_s T_{m,s}(\tau)$ exists.
$\Rightarrow \lim_s T_{e+1,s}(\emptyset)\downarrow$.

Proof: First, by Lemma A, $\forall \tau$, $\lim_s T^{rec}_{e,s}(\tau)$ exists. Now suppose to the contrary that $\lim_s T_{e+1,s}(\emptyset)\not\downarrow$.

(i) We may still define $\lim_s \beta_s = B$.

We claim that $T_e^{rec}(0^m)\downarrow\ \forall m$, and given any m, $\exists t$ such that $\forall s > t\ \beta_s \supseteq T_e^{rec}(0^m)$. This will give us $\lim_s \beta_s = \bigcup_{m\geq 0} T_e^{rec}(0^m)$.

It is clearly true for $m = 1$. Suppose it is true for m. Since $\lim_s T_{e+1,s}(\emptyset) \neq T_e^{rec}(0^m)$ we must define $T_e^{rec}(0^m * 0)(0^m * 1)$ and erect a marker for $T_e^{rec}(0^m * 0)(0^m * 1)$ infinitely often. Since we have a marker infinitely often we must have a permanent marker, say at $\forall s > s_1$. Then $\forall s > s_1$, $T_{e+1,s}(\emptyset) \supseteq T_{e,s}^{rec}(0^{m+1})$. So $\forall s > s_1$, $\beta_s \supseteq T_{e,s}^{rec}(0^{m+1}) = T_e^{rec}(0^{m+1})$. So by induction (i) is proved. □

(ii) We may define $T_m^* \ \forall m \leq e$.

We may define T_m^* <u>exactly</u> as in Lemma 3 of <u>m</u> < <u>0</u>' (page 35).

We claim that (1)-(6), except for (2), follow exactly as in Lemma 3 of <u>m</u> < <u>0</u>'. Simply note that the "extra" dummy strings on $T_{r,s}^{rec}$, $\forall r$, cause no interference.

We prove

(2) T_{m+1}^* is either an $m+1$-splitting tree

or $\exists \beta \subset B$ such that no pair of strings lying above β on T_{e+1}^* $m+1$-split β.

<u>Proof</u>: If we define T_{m+1}^* as an $m+1$-splitting tree by case (iii), then (2) is immediate. So suppose $T_{m+1}^* = F_{\pi(m+1)}(T_m^*)$.

If we used case (i) to define T^*_{m+1} then we have an end string on T_{m+1}, $T_{m+1}(t) \subset \pi(m+1)$ as in $\underline{m} < \underline{0}'$. Suppose (2) fails. Then we must have $m+1$-splittings lying on T^*_{m+1} above arbitrarily long beginnings of B. Since we do not have $T_{m+1}(\tau * 0)(\tau * 1)\downarrow$, no boundary strings ever intervene between $T_{m+1}(\tau)$ and these splittings. Otherwise we would use Case III to permanently define $T_{m+1}(\tau * 0)(\tau * 1)$. So we must define infinitely many permanent potential extensions of $T_{m+1}(\tau)$, with associated markers $(m+1, r_n, s_n)_{n=1}^{\infty}$, $r_n < r_{n+1}$, $s_n < s_{n+1}$. We may obtain this sequence recursively: we define $(m+1, r_1, s_1)$ to be the first marker associated with any potential extension lying on T^*_m with $s_1 > s(m+1)$; and $(m+1, r_{n+1}, s_{n+1})$ to be the first marker associated with any potential extension lying on T^*_m with $s_{n+1} > s_n$. Since $\forall s > s(m+1)$, $\beta_s \supseteq \pi(m+1)$, these are the permanent potential extensions of $T_{m+1}(\tau)$. The sequence is recursive since T^*_m is partial recursive. Since we can never use Case II to obtain $T_{m+1}(\tau * 0)(\tau * 1)$ we must have for each n, $\forall s \geqslant s_n$, $f(s) \geqslant r_n$. This implies that A is recursive: to compute $A(x)$ let $n(x)$ be minimal such that $r_{n(x)} > x$; then $x \in A \leftrightarrow x = f(s)$ some $s < s_{n(x)}$. This is a contradiction.

T^*_{m+1} defined by case (ii) is similar. □

(iii) We arrive at a contradiction using (ii).

Recall that we have supposed $\lim_s T_{m+1,s}(\emptyset)\not\downarrow$. We have shown that $\lim_s \beta_s = B$ exists and that T^*_m $m \leqslant e$, have the appropriate properties.

As we showed above, every branching $T_e^{rec}(0^m * 0)(0^m * 1)$ must have a permanent marker attached to it, say $(e\ \text{rec}, r_m, s_m)$, where $r_m < r_{m+1}$, $s_m < s_{m+1}$. As in part (ii) we can obtain this sequence recursively. None of these markers is ever permanently valid, for otherwise some t_0, $\forall s > t_0$, $T_{e+1,s}(\emptyset) = T_{e,s}^{rec}(0^m * i)$ some m. So $\forall m$, $\forall s > s_m$, $f(s) > r_m$. As in (ii) above, this implies that A is recursive, contradiction. □

Lemma D) B is not recursive.

<u>Proof</u>: Suppose $B = \phi_{e+1}$. Choose e minimal. We must have no valid marker on T_e^{rec}. Since ϕ_{e+1} is total we must erect markers for $T_e^{rec}(0^m * 0)(0^m * 1)$ $\forall m$. But then given m, $\exists t\ \forall s > t$, $T_{e+1,s}(\emptyset) \supseteq T_e^{rec}(0^m * 0)$, a contradiction on $\lim_s T_{e+1,s}(\emptyset)\downarrow$. □

Lemma E) $B \leqslant_T A$.

<u>Proof</u>: Using A we give a computation procedure for B.

Let $s(x) = \mu s(\forall t \geqslant s,\ f(t) \geqslant x)$. Then $s(x) \leqslant_T A$.

Let $t(x) = \mu t(t \geqslant s(x),\ \beta_t(x)\downarrow)$. Then $t(x) \leqslant_T A$.

We claim that $B(x) = \beta_{t(x)}(x)$.

If $s \geqslant s(x)$ we may not apply permitting below x at stage s. However, we may have $\beta_s(x)\downarrow$, $\beta_{s+1}(x)\not\downarrow$. (See diagram V-6). But in any case, $\forall s \geqslant t(x)$, $\beta_{t(x)}[x] \subseteq$ retraced path at s. So $\forall s \geqslant t(x)$, $\beta_t(x)[x]$ is compatible with β_s. So $\lim_s \beta_s[x] = \beta_{s(x)}[x]$. $\square$

This concludes the proof of $\underline{m} < \underline{a}$ r.e.

CHAPTER VI: A MINIMAL DEGREE such that $\underline{m} \cup \underline{a}$ r.e. = $\underline{0}'$

Sections

* * *

In this Chapter we prove that given $\underline{0} < \underline{a}$ r.e. there is a minimal degree $\underline{m} < \underline{0}'$ such that $\underline{m} \cup \underline{a} = \underline{0}'$. Here "$\cup$" indicates the "join" in the upper semi-lattice of degrees. As a corollary we have that every r.e. degree has a complement in the upper semi-lattice of degrees $\leqslant \underline{0}'$. This contrasts with a theorem of Lachlan (4) that we cannot find a complement for $\underline{a}$ in the upper semi-lattice of r.e. degrees.

In "The Join Technique" we present in rudimentary form a technique for finding $\underline{b}$ such that $\underline{b} \cup \underline{a} = \underline{0}'$. This method is due to R.W. Robinson (6). The construction of $\underline{m} \cup \underline{a}$ r.e. = $\underline{0}'$ is then just a wedding of this method to the permitting techniques we established in $\underline{m} < \underline{a}$ r.e. We will assume the reader is familiar with Chapter V. The plan of proof and several proofs from that chapter will carry over with very little modification.

In "Further Topics and Remarks" we survey what else is known about degrees complementary below $\underline{0}'$, and pose some questions for the reader.

In this chapter we will assume that $\underline{a}$ is r.e. and $\underline{0} < \underline{a} < \underline{0}'$. We may assume $\underline{a} \neq \underline{0}'$ since the theorem follows from $\underline{m} < \underline{0}'$ in that case. We will have:

$$A \in \underline{a},\ K \in \underline{0}',\ f(\mathbb{N}) = A,\ g(\mathbb{N}) = K,\ f,g \quad \text{1-1, recursive.}$$

The Join Technique

We illustrate a technique which enables us to construct $\underline{b} < \underline{0}'$ such that $\underline{b} \cup \underline{a} = \underline{0}'$.

Suppose we have A, f, and K, g as above. We want to make whatever B we construct satisfy $A \oplus B \geq_T K$. Here $A \oplus B = \{2x, 2y+1 : x \in A, y \in B\}$. Since $\underline{b} < \underline{0}'$ we will then have $\underline{a} \cup \underline{b} = \underline{0}'$.

We will construct B by recursive stages, $\lim_s \beta_s = B$. We will want B to settle down slowly enough so that, in conjunction with the rate at which A settles down we measure the rate at which K settles down. To do this we will construct with β_s an auxiliary recursive function $t(e,s)$, ensuring that $\lim_s t(e,s)\downarrow \; \forall e$. We will let

$$r(e) = \mu s(s > r(e-1) \text{ and } \forall x \leq t(e,s)\; A^s(x) = A(x) \wedge \beta_s(x) = B(x)),$$

where $A^s = \{f(x) : x \leq s\}$. We need $\lim_s t(e,s)\downarrow$ to have $r(e)\downarrow$. $r(e) \leq_T A \oplus B$. If we can arrange our approximation $\{\beta_s\}_{s\geq 0}$ so that

$$e \in K \leftrightarrow e \in K^{r(e)}$$

then we will have $K \leq_T A \oplus B$.

Since K is r.e. we will only have to make certain that $e \in K \rightarrow e \in K^{r(e)}$. If $e \notin K$ then $e \notin K^{r(e)}$ $\forall e$.

The reader may ask why we use $t(e,s)$ rather than just measuring the rate at which $A^s[e] \wedge \beta_s[e]$ first read right. It will soon become clear why that is too restrictive.

Schematically, then, this is what we do to our approximation $\{\beta_s\}_{s\geq 0}$.

stage s: At the end of stage s, $t(e,s)\downarrow$.

Let r be minimal such that $A^r[t(e,s-1)] = A^s[t(e,s-1)]$.

Since A is r.e. $\rightarrow A^u[t(e,s-1)] = A^r[t(e,s-1)]\ \forall u$,

$r \leq u \leq s$.

Suppose $g(s) = e$.

Set $\beta_s[t(e,s-1)] \neq \beta_u[t(e,s-1)]\ \forall u,\ r \leq u < s$,

and set $t(e,s) = t(e,s-1)$.

If we require $\forall w \geq s+1,\ \beta_w[t(e,s-1)] = \beta_s[t(e,s-1)]$, then $r(e) \geq s+1$.

We call this a "switch for g"; we change our approximation to B because $g(s) = e$.

This is easy to arrange. The difficulty is balancing this "switch for g" against whatever other requirements we may have. Suppose we have at stage $w > s$ a string, or pair of strings which we would like to use to satisfy some requirement. And suppose these would force $B[t(e,s)] = \beta_r[t(e,s)]$. Then they would force B to level s to settle down too quickly -- at $r < s$, so that $r(e) < s$. What do we do?

We will use this string or strings if f permits us to, by $f(v) < t(e,s)$ some $v \geq w$. For if A hasn't settled down to level $t(e,s)$ at stage s it would be all right for $B[t(e,s)] = \beta_r[t(e,s)]$. We would have $r(e) \geq v > s$ due to A.

But we can't wait for $g(s) = e$ before trying to satisfy other requirements. For infinitely many e, $g(s) \neq e, \forall s$. So schematically again,

stage s: $t(e,s-1)\downarrow$. r minimal such that $A^r[t(e,s)] = A^s[t(e,s)]$. We see a string or pair of strings which we would like to use to satisfy some requirement of priority higher than doing a "switch for g" at e. And this string would fix B to level $t(e,s-1)$. (Remember it can fix it badly by making $B[t(e,s-1)] = \beta_s[t(e,s-1)]$. Then if $A^r[t(e,s-1)] = A[t(e,s-1)]$ we would have $r(e) \leq s-1$ even if $g(t) = e$ some $t \geq s$.)

We set $\beta_s[t(e,s-1)] \neq \beta_u[t(e,s-1)]$ $\forall u$, $r \leq u \leq s$, just as we did above (except s not $s+1$). Again, we'll require $\beta_w[t(e,s-1)] = \beta_s[t(e,s-1)]$ $\forall w > s$ unless f permits us to do otherwise. Only we'll make $t(e,s) \gneq t(e,s-1)$.

We cannot calculate $r(e)$ from $t(e,s-1)$ before stage s because B and A couldn't have settled down to level $t(e,s-1)$ before stage s. But at stage s we have made sure that we can't calculate $r(e)$ from $t(e,s-1)$ by making $t(e,s) \gneq t(e,s-1)$. We will have to pick $t(e,s)$ so that we can make a switch for g if $g(t) = e$ some $t \geq s$.

As we explained before, we will use the string or strings we have picked out only if some $v > s$, $f(v) \leq t(e,s-1)$.

Now we ask for a string or strings which satisfy this requirement which lie above $\beta_s[t(e,s-1)]$.

Two basic questions arise: one, how do we assure that we do not push up $t(e,s)$ infinitely often? And two, how do we know that we will be able to satisfy every requirement? These are closely intertwined. We will only push up $t(e,s)$ if we are trying to satisfy a requirement of priority higher than "switching for g at e."

For each of these requirements either we will have only finitely many chances to satisfy it in which case we will push up $t(e,s)$ only finitely often for it. Or we will have infinitely many chances to satisfy it, in which case by the usual permitting argument, since A is not recursive, f will eventually permit us to satisfy it, and that requirement will stop pushing up $t(e,s)$.

Motivation

How do we incorporate the join technique into our $\underline{m} < \underline{0}'$ construction? First, let's consider where we'll "switch for g at e." We will set up a place between $T_{e,s}(\emptyset)$ and $T_{e+1,s}(\emptyset)$ by introducing a tree $T^{join}_{e,s}$.

VI-1

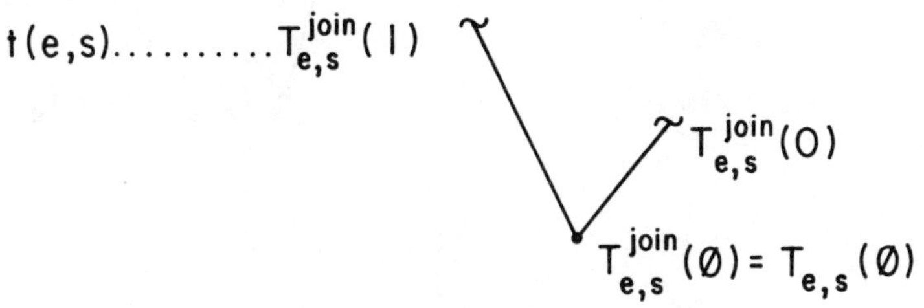

$T^{join}_{e,s}$ will be very similar to $T^{rec}_{e,s}$ of $\underline{m} < \underline{a}$ r.e.: it will consist only of least dummy extensions, and its purpose will be the placement of $T_{e+1,s}(\emptyset)$. We will make $lh(T^{join}_{e,s}(1)) \geq lh(T^{join}_{e,s}(0))$. The reason for this will become clear later. We will set $t(e,s) = lh(T^{join}_{e,s}(1))$.

As an aside note that since we will have $\underline{a} \cup \underline{b} = \underline{0}'$ and $\underline{a} < \underline{0}'$ we will automatically have $\underline{b} \neq \underline{0}$. So we won't need $T^{rec}_{e,s}$ in this construction. $T_{e,s}$ will be concerned only with e-splittings.

This first branching we have drawn will be the initial place where we will switch for g if $g(s) = e$. So we need $\beta_u \supseteq T^{join}_{e,u}(0)$ for $\forall u$, $r \leq u \leq s$, where r is minimal such that $A^r[t(e,s)] = A^s[t(e,s)]$. This we can accomplish by making $T_{e+1,u}(\emptyset) = T^{join}_{e,u}(0)$.

If $g(s) = e$ we will make $\beta_s \supseteq T^{join}_{e,s}(1)$ by setting

$T_{e+1,s}(\emptyset) = T^{join}_{e,s}(1)$. We will keep $T_{e+1,w}(\emptyset) = T^{join}_{e,w}(1) = T^{join}_{e,s}(1)$, $w > s$, unless f permits us to make a change.

<u>VI-2</u> <u>stage s</u> and $g(s) = e$

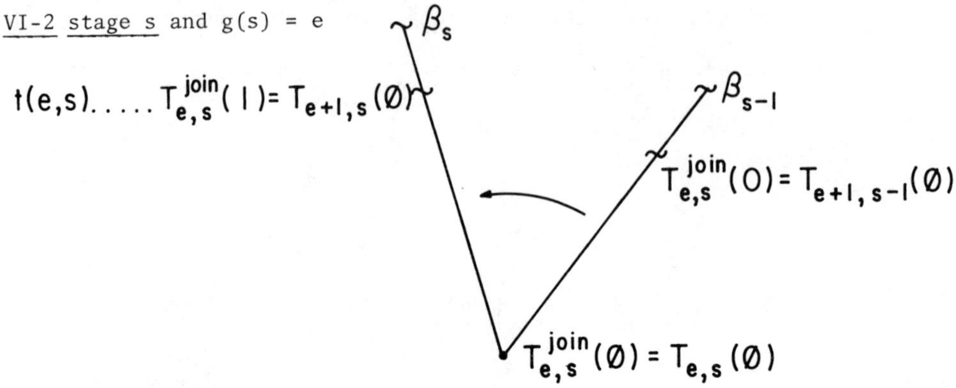

This takes care of the easy part, making B settle down slowly enough. Now how do we do the hard part, satisfying the other requirements? In this case the other requirements are: putting up e'-splittings when necessary for $e' \leq e$.

First recall that we can't wait for $g(s) = e$ before e'-splitting. So when we e'-split, or attempt to, we'll have to make certain that we keep available a place to do a switch for g at e.

So suppose we see a splitting for $e' \leq e$ which we would like to use and which would cancel the branching of diagram IV-1, $T^{join}_{e,s}(0)(1)$.

VI-3

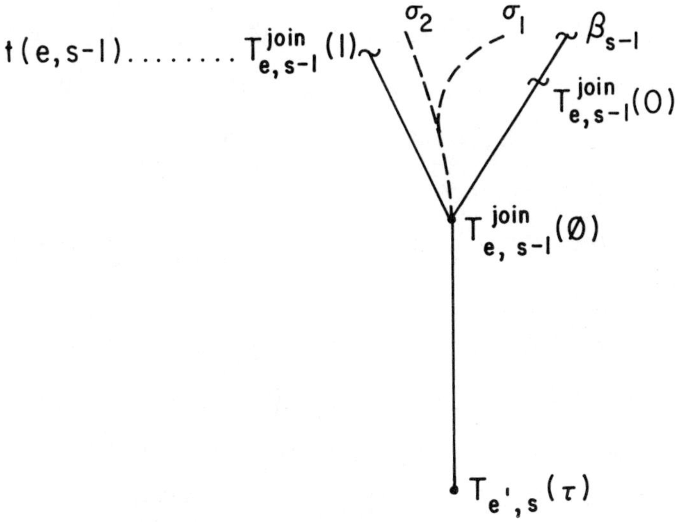

We would like to have $T_{e',s}(\tau * 0)(\tau * 1) = \sigma_1, \sigma_2$. Since we're working on $T_{e',s}$ we haven't yet defined $T^{join}_{e,s}(0)(1)$. But we know where $T^{join}_{e,s-1}(0)(1)$ are, and it just could be that we have:

VI-4

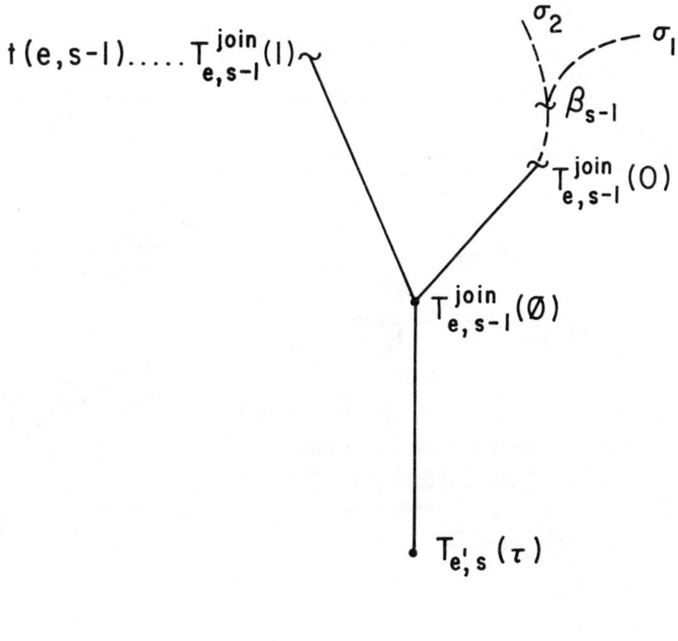

That is the problem. For if we use σ_1, σ_2 then we force B to settle down to level $t(e,s-1)$. And $B[t(e,s-1)] = \beta_{s-1}[t(e,s-1)]$ would mean that $r(e) \leq s - 1$ even though we might have $g(t) = e, t \geq s$.

We can't just go ahead and use σ_1, σ_2. We must wait for f to permit us to, by $f(v) < lh(T^{join}_{e,s-1}(\emptyset))$, $v \geq s$. So we'll erect a marker $(e', lh(T^{join}_{e,s-1}(\emptyset)), s)$. (σ_1, σ_2) is now a potential extension of $T_{e',s}(\tau)$.

Where shall we look for another e'-splitting in case we can't use this one? Above $T^{join}_{e,s-1}(0)$? No, for then, just as here, we would need $f(v) \leq T^{join}_{e,s-1}(\emptyset)$. We need to <u>force</u> β_s above $T^{join}_{e,s-1}(1)$. Then we can look for e'-splittings above $T^{join}_{e,s-1}(1)$.

We set $T^{join}_{e,s}(0)(1) = T^{join}_{e,s-1}(0)(1)$ and make $T_{e+1,s}(\emptyset) \supseteq T^{join}_{e,s}(1)$. We will keep $\beta_w \supseteq T^{join}_{e,w}(1) = T^{join}_{e,s}(1)$, $w > s$, unless f permits us otherwise.

But then we'll need a new place to do our switch for g at e if necessary. We have "used up" $T^{join}_{e,s}(0)(1)$ by having used both sides of this branching as beginnings of B. We'll define $T^{join}_{e,s}(1 * 0)(1 * 1)$.

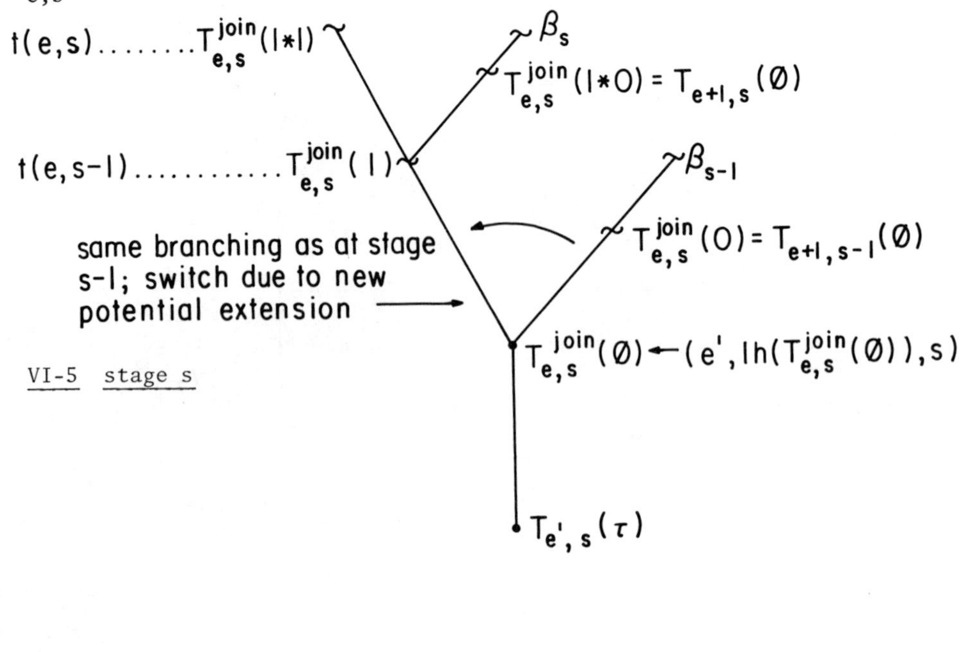

<u>VI-5</u> <u>stage s</u>

We will put $T_{e+1,s}(\emptyset) = T^{join}_{e,s}(1 * 0)$. And we will set $t(e,s) = lh(T^{join}_{e,s}(1 * 1)) \geqslant lh(T^{join}_{e,s}(1 * 0))$. Can you now see why we need $lh(T^{join}_{e,s}(1)) \geqslant lh(T^{join}_{e,s}(0))$? So $t(e,s) > t(e,s-1)$.

We will keep $T_{e+1,w}(\emptyset) = T^{join}_{e,w}(1 * 0)$ as long as, $w \geqslant s$, $T^{join}_{e,w}(1 * 0)(1 * 1)\downarrow$ unless we need to switch for g at e or unless another e'-splitting shows up which forces β_w above $T^{join}_{e,w}(1 * 1)$. So $T^{join}_{e,w}(1 * 0)(1 * 1)$ will be a suitable place to do our switch for g, at least until another potential e'-splitting shows up. If that happens, we'll force β_w above $T^{join}_{e,w}(1 * 1)$ and define $T^{join}_{e,w}(1^2 * 0)(1^2 * 1)$ as a suitable branching to do our switch for g at e if $g(t) = e$, $t > w$. ($1^m = \underbrace{1*\ldots*1}_{m\text{-times}}$, $1^0 = \emptyset$).

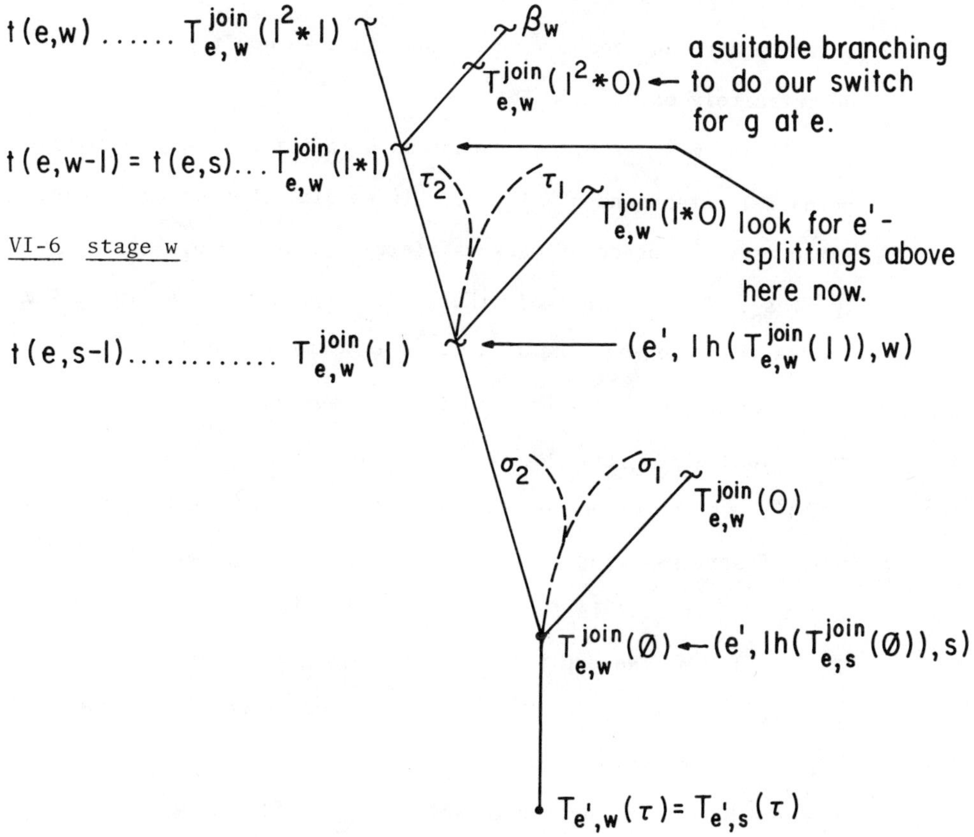

VI-6 stage w

As we explained in the previous section, by pushing up $t(e,s)$ in diagram VI-5, we insure that $r(e)$ won't be calculated from $t(e,s-1)$, even though $B[t(e,s-1)] = \beta_s[t(e,s-1)]$ is possible.

As an aside, consider how this differs from $\underline{m} < \underline{a}$ r.e. It's just "backwards." In $\underline{m} < \underline{a}$ r.e. we were avoiding splittings which might make us settle down too slowly. So we forced β_s above $T^{rec}_{e,s}(0)$ and tried to satisfy requirements above $T^{rec}_{e,s}(0)$. Here we are worried about splittings which could make us settle down too quickly. So we change our approximation by forcing β_s above $T^{join}_{e,s}(1)$ and try to split above $T^{join}_{e,s}(1)$.

Now let's consider those two sticky problems we posed at the end of the previous section: how do we know that we can satisfy every requirement and how do we know that we don't push $t(e,s)$ up infinitely often?

As in $\underline{m} < \underline{a}$ r.e. we cannot have infinitely many permanent potential extensions of $T_{e'}(\tau)$. If we did, then we would have a recursive sequence of markers (e,r_n,s_n) and $\forall n$, $f(t) > r_n$, $\forall t > s_n$. This would imply that A is recursive. As in $\underline{m} < \underline{a}$ r.e., to see that the sequence is recursive we must pass to $T^*_{e'-1}$.

Because we cannot have infinitely many permanent potential extensions of $T_{e'}(\tau)$ we will have that $T^*_{e'}$ satisfies:

(2) $T^*_{e'}$ is either an e'-splitting tree, or $\exists \beta \subset B$ such that no pair of strings lying above β on $T^*_{e'}$ e'-split β. But unlike $\underline{m} < \underline{a}$ r.e. it is this fact which also implies that we do not define $T^{join}_e(1^m)$ $\forall m$. We explain.

We mustn't define $T^{join}_e(1^m) \forall m$, for if we did we would push $t(e,s)$ up infinitely often. Or equivalently, $T_{e+1,s}(\emptyset)$ would not reach a limit. We prefer to watch $T_{e+1,s}(\emptyset)$ rather than $t(e,s)$ for our limit argument. Defining a new branching on $T_{e',s}$ can affect $T_{e+1,s}(\emptyset)$ and the limit process only as it does in $\underline{m} < \underline{0}'$. Either we stop defining branchings on $T_{e',s}$, and then we have only finitely many permanent potential extensions of $T_{e'}(\tau)$, or else we continue to define branchings. The latter cannot cause $T_{e+1}(\emptyset)\uparrow$ for the same reason it doesn't in $\underline{m} < \underline{0}'$ -- it just changes the least dummy pair which must eventually settle down. If $T_{e'}(\tau)$ has finitely many permanent potential extensions this won't push up $T_{e+1,s}(\emptyset)$ infinitely often. But is it possible that for infinitely many δ, $T_{e'}(\delta)$ has finitely many permanent potential extensions? That, indeed, could push up $t(e,s)$ infinitely often. The answer though, is no, for in that case we will be able to claim that the permanent potential extensions of $T_{e'}(\delta)$ for all sufficiently long δ would also be suitable as potential extensions of some fixed $T_{e'}(\tau) \supset T^*_{e'-1}(\emptyset)$. This will be because no boundary strings will intervene, as the permanent potential extensions all lie along the path of B. So $T_{e'}(\tau)$ would have infinitely many permanent potential extensions which it does not. In this way we'll be able to prove that $\lim_s T_{e+1,s}(\emptyset)\downarrow$ and hence $\lim_s t(e,s)\downarrow$.

By focussing on $T^{join}_{e,s}(0)(1)$ in our discussion we have blurred one point which we wish to clear now. Consider

VI-7

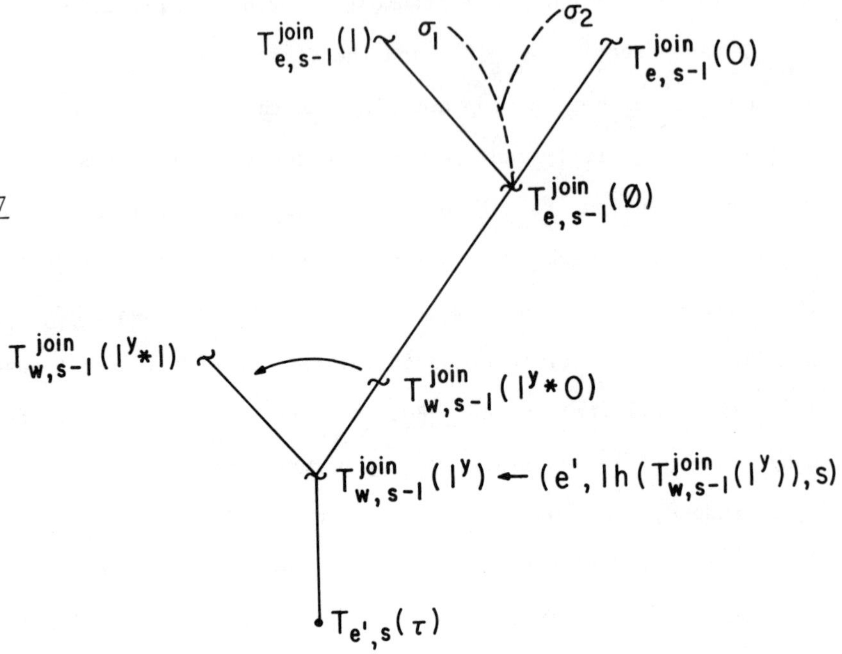

When we spot σ_1, σ_2 we want to force β_s above the first branching $T^{join}_{w,s-1}(1^y * 0)(1^y * 1)$ which has $\beta_{s-1} \supseteq T^{join}_{w,s-1}(1^y * 0)$. In diagram VI-7 that is not $T^{join}_{e,s-1}(0)(1)$. When we force $\beta_s \supseteq T^{join}_{w,s}(1^y * 1)$ we will cancel $T^{join}_{e,s-1}(0)(1)$.

Note two things: $T_{e'}$ will, by its potential extensions, force β_s above the left hand side of all the $T^{join}_{w,s}$ branchings above it consecutively until it causes no more forcings. And cancelling $T^{join}_{e,s-1}(0)(1)$ is all right. We can get β_t back above $T^{join}_{e,s-1}(0)$, $t > s$, only if we have permitting taking place below $lh(T^{join}_{w,s-1}(1^y))$. This is important since we don't accidentally want to get $\beta_t \supset T^{join}_{e,s-1}(0)$ $\forall t > u$, some u, without permitting taking place. For then we would have $r(e) \leqslant s-1$ since B would have read right to level $t(e,s-1)$ at stage s-1.

To summarize: we hold onto our path $T^{join}_{e,s}(1^m * 0)$, so that we may make a switch to $T^{join}_{e,s}(1^m * 1)$ if $g(t) = e$, $t \geqslant s$. We switch from this path to $T^{join}_{e,s}(1^m * 1)$ if g makes us, or if we see a nice splitting. Then $T^{join}_{e,s}(1^m * 1)$ becomes the retraced path we want, until such time as f permits us to change from it. We need to keep pushing the retraced path up via $T^{join}_{e,s}(1^m * 1)$ to eventually find splittings for $e' \leq e$. $T_{e'}$ may injure $t(w,s)$ $\forall w \geq e'$, but it can do so, for each w, at most finitely often.

This completes the motivation. Join us in the construction, if you will, but don't jump to any conclusions.

<u>Construction of $\underline{m} \cup \underline{a}$ r.e. $= \underline{0}'$</u>

Let $T_{0,s}$ = identity tree to level s.

At stage s all strings are chosen from $T_{0,s}$.

<u>$T_{e,s}$</u>, $e > 0$: $T_{e,s}(\emptyset)$: Let m be maximal such that $T^{join}_{e-1,s}(1^m * 0)(1^m * 1)\downarrow$

If β_s is <u>forced</u> above $T^{join}_{e-1,s}(1^m * 1)$ set

$T_{e,s}(\emptyset) = T^{join}_{e-1,s}(1^m * 1)$.

If not, set $T_{e,s}(\emptyset) = T^{join}_{e-1,s}(1^m * 0)$.

(Set $T_{1,s}(\emptyset) = T_{0,s}(0)$).

We continue only if $T_{e,s}(\tau)\downarrow = T_{e,s-1}(\tau)$.

Case I: $T_{e,s-1}(\tau * 0)(\tau * 1)\downarrow$ and compatible with $T_{r,s}\ \forall r < e$

and [(i) $\tau = \emptyset$] ← delete

or (ii) $T_{e,s-1}(\tau * 0)(\tau * 1)\downarrow$ originally by Case II and (b) of Case II (potential) still applies,

or (iii) $\nexists\, \sigma_1, \sigma_2$ as in Case II; or $\beta_{s-1} \not\supseteq T_{e,s}(\tau)$.

Then set $T_{e,s}(\tau * 0)(\tau * 1) = T_{e,s-1}(\tau * 0)(\tau * 1)$.

We continue only if $T_{e,s}(\tau) \subseteq \beta_{s-1}$.

Case II(potential): If $T_{e,s}(\tau * 0)(\tau * 1)\downarrow$ it is not originally by Case II; and $\exists\, \sigma_1, \sigma_2$ such that

(a) σ_1, σ_2 are compatible with $T_{r,s}\ \forall r \leq e$.

(b) Every $\sigma' \subset \sigma_1$ or $\sigma' \subset \sigma_2$ which is a boundary string for some $r < e$ satisfies $\sigma' \subsetneqq T_{e,s}(\tau)$.

(c) σ_1, σ_2 e-split $T_{e,s}(\tau)$ at s

(d) If the last appointed current potential

extension of $T_{e,s}(\tau)$ has marker (e,r,t) then $\exists\, w,y$ such that $T^{join}_{w,s-1}(1^y * 0)(1^y * 1)\downarrow$ and $lh(T^{join}_{w,s-1}(1^y)) > r$.

(If there is no such marker require only that $\exists\, w,y$ such that $T^{join}_{w,s-1}(1^y) \supsetneq T_{e,s}(\tau)$).

Choosing the first such w,y then $\sigma_1 \wedge \sigma_2 \supseteq T^{join}_{w,s-1}(1^y)$.

(See diagram VI-7 and explanation page 110).

Then we appoint the least such (σ_1,σ_2) a potential extension of $T_{e,s}(\tau)$, and associate with it the marker $(e, lh(T^{join}_{w,s-1}(1^y)), s)$.

If β_s is not already forced above $T^{join}_{w,s-1}(1^y)$ it now is.

(σ_1,σ_2) is a current potential extension of $T_{e,u}(\tau)$, $u \geq s$, until it is cancelled. It is not cancelled only if $T_{e,u}(\tau) = T_{e,s}(\tau)$ and (a), (b), (c) of Case II(potential) continue to hold with s replaced by u, and $T_{e,u}(\tau) \subseteq \beta_{u-1}$. If any of these conditions fail, or if $T_{e,u}(\tau * 0)(\tau * 1)\downarrow$ by Case II, then (σ_1,σ_2) is immediately cancelled.

Case II: Case I does not apply and $\exists$ a potential extension (σ_1,σ_2) of $T_{e,s}(\tau)$ with marker (e,r,t) and $f(s) < r$.

Then set $T_{e,s}(\tau * 0)(\tau * 1)$ = least such σ_1,σ_2.

This is an <u>application of permitting</u> at r.

Case III: Cases I, II, and II (potential) do not apply

and [(i) $\tau = \emptyset$] ← delete

or (ii) $\exists\, \sigma_1,\sigma_2$ as in Case II(potential) except that (b) fails.

Set $T_{e,s}(\tau * 0)(\tau * 1)$ = least dummy extension of $T_{e,s}(\tau)$.

* * *

<u>$T^{join}_{e,s}$</u>, $e > 0$: $T^{join}_{e,s}(\emptyset) = T_{e,s}(\emptyset)$.

$T^{join}_{e,s}(0)(1) = \sigma_1,\sigma_2$ such that (σ_1,σ_2) is the least dummy extension of $T^{join}_{e,s}(\emptyset)$ and $lh(\sigma_2) \geq lh(\sigma_1)$.

Suppose $T^{join}_{e,s}(1^m * 0)(1^m * 1)\!\downarrow$, $m \geq 0$.

We proceed by the appropriate case below:

(1) $T^{join}_{e,s}(1^m * 0)(1^m * 1) = T^{join}_{e,s-1}(1^m * 0)(1^m * 1)$ and β_{s-1} or β_s is forced above $T^{join}_{e,s-1}(1^m * 1)$

β_s is now <u>forced above</u> $T^{join}_{e,s}(1^m * 1)$.

(a) If $t(e,u)$ was <u>fixed</u> for g some $u \leq s-1$ ($g(u) = e$) and $t(e,u) = t(e,s-1) = lh(T^{join}_{e,s-1}(1^m * 1))$ then set $t(e,s) = t(e,s-1)$. Make no further extensions on $T^{join}_{e,s}$.

(b) Not (a). Then define

$T^{join}_{e,s}(1^{m+1} * 0)(1^{m+1} * 1) = \tau_1, \tau_2$ such that (τ_1, τ_2) is the least dummy extension of $T^{join}_{e,s}(1^{m+1})$ and $lh(\tau_2) \geqslant lh(\tau_1)$.

(2) Not (1). That is $T^{join}_{e,s}(1^m * 0)(1^m * 1) \not\equiv T^{join}_{e,s-1}(1^m * 0)(1^m * 1)$ or, neither β_s nor β_{s-1} is forced above $T^{join}_{e,s-1}(1^m * 1)$. Make no further extensions on $T^{join}_{e,s}$. Set $t(e,s) = lh(T^{join}_{e,s}(1^m * 1))$.

<u>Fixing a **marker** for g</u>:

If $g(s) = e$, see if $\exists m$ such that $T^{join}_{e,s}(1^m * 0)(1^m * 1)\downarrow$ and β_s is not forced above $T^{join}_{e,s}(1^m * 1)$. If there is none do nothing.

If there is one then $t(e,s) = lh(T^{join}_{e,s}(1^m * 1)$. β_s is now <u>forced above</u> $T^{join}_{e,s}(1^m * 1)$. And $t(e,s)$ is <u>fixed</u> for g.

Proceed to $T_{e+1,s}$.

* * *

At the end of stage s set

$$m_s = \max m(T_{m,s}(\emptyset)\downarrow)$$

and

$$\beta_s = T_{m_s,s}(\emptyset).$$

Plan of Proof

The plan of proof concerning the T_e is, in its essentials, the same as in $\underline{m} < \underline{a}$ r.e. Once we can show that $\lim_s T_{e,s}(\emptyset)\downarrow$, $\forall e$, we will be done. Along the way we prove that T^*_e, $\forall e$, satisfy the appropriate properties ((1)-(6) of $\underline{m} < \underline{0}'$). In outline, then, the plan is:

Lemma A) (1) $\forall m \leqslant e$, $\forall \tau$, $\lim_s T_{m,s}(\tau)$ exists and $\lim_s T_{e+1,s}(\emptyset)\downarrow$

$\Rightarrow$ $\forall \tau$, $\lim_s T_{e+1,s}(\tau)$ exists.

- As in $\underline{m} < \underline{a}$ r.e. we do not expand on this.

(2) $\forall m \leqslant e$, $\forall \tau$, $\lim_s T_{m,s}(\tau)$ exists $\Rightarrow \forall \tau$ $\lim_s T^{join}_{e,s}(\tau)$ exists.

- We will prove this.

Lemma B) $\forall m \leqslant e$, $\forall \tau$, $\lim_s T_{m,s}(\tau)$ exists $\Rightarrow \lim_s T_{e+1,s}(\emptyset)\downarrow$.

(i) If it were not we could still define $B = \lim_s \beta_s$.

- The proof is similar to $\underline{m} < \underline{a}$ r.e.

(ii) Since we have B, we may define T^*_m $\forall m \leqslant e$ as in Lemma 3 of $\underline{m} < \underline{0}'$. Properties (1)-(6) are immediate except for (2) (splittings always available) which we will prove.

(iii) We arrive at a contradiction using (i) and (ii).

- The proof of this is significantly different from $\underline{m} < \underline{a}$ r.e.

Corollary: $\lim_s t(e,s)\downarrow$.

Lemma C) $\underline{a} \cup \underline{b} = \underline{0}'$.

- By Lemmas (A) and (B), B is defined. We

will give a computation procedure as outlined in "The Join Technique."

Lemma D): $\underline{b}$ is minimal.

- Since $\underline{a} < \underline{0}'$, $\underline{b} \neq \underline{0}$ from Lemma C). The T_e^*, $\forall e$, have the appropriate properties to show that b is minimal (from Lemma(B) - repeat that argument using the real B). We do not expand any further on this.

Proof

Lemma A) (2) $\forall m \leq e$, $\forall \tau$, $\lim_s T_{m,s}(\tau)$ exists $\Rightarrow$ $\forall \tau$, $\lim_s T^{join}_{e,s}(\tau)$ exists.

Proof: $\lim_s T^{join}_{e,s}(\emptyset) = T_e(\emptyset)$.

$\lim_s T^{join}_{e,s}(0)(1)\downarrow$ as this branching is just the least dummy extension of $T^{join}_{e,s}(\emptyset)$.

Assume $\lim_s T^{join}_{e,s}(1^m * 0)(1^m * 1)\downarrow$. Let s_0 be such that $T^{join}_{e,s_0}(1^m * 0)(1^m * 1)$ have settled down. See if $\exists s_1 \geq s_0$ such that β_{s_1} is forced above $T^{join}_{e,s_1}(1^m * 1)$. If there is none then we have $T^{join}_{e,s}(1^{m+1} * 0)(1^{m+1} * 1)\not\downarrow$ $\forall s \geq s_0$. So $\lim_s T^{join}_{e,s}(1^{m+1} * 0)(1^{m+1} * 1)$ exists.

If there is such an s, then $\forall s > s_1$, β_s is forced above $T^{join}_{e,s}(1^m * 1)$. This is because the only way we can have β_s not forced above $T^{join}_{e,s}(1^m * 1)$ is if we cancel $T^{join}_{e,s-1}(1^m * 0)(1^m * 1)$, which we do not. So $\forall s > s_1$, $T^{join}_{e,s}(1^{m+1} * 0)(1^{m+1} * 1)\downarrow$.

Let $s_2 \geq s_1$ be such that $\forall s > s_2$ the least dummy extension of $T^{join}_{e,s}(1^{m+1})$ is well defined, say (τ_1, τ_2), $lh(\tau_2) \geq lh(\tau_1)$. Then $\forall s > s_2$, $T^{join}_{e,s}(1^{m+1} * 0)(1^{m+1} * 1) = \tau_1, \tau_2$. So $\lim_s T^{join}_{e,s}(1^{m+1} * 0)(1^{m+1} * 1)\downarrow$. □

Lemma B) $\forall m \leq e$, $\forall \tau$, $\lim_s T_{m,s}(\tau)$ exists $\Rightarrow$ $\lim_s T_{e,s}(\emptyset)\downarrow$.

Proof: By Lemma A) $\forall \tau$, $\lim_s T^{join}_{e,s}(\tau)$ exists.

Suppose to the contrary that $\lim_s T_{e+1,s}(\emptyset)\not\downarrow$.

(i) We could still define $\lim_s \beta_s = B$.

Proof: If there were a maximal m such that $T^{join}_e(1^m * 0)(1^m * 1)\downarrow$ then $\forall s > s_0$, some

s_0, β_s is not forced above $T^{join}_{e,s_0}(1^m * 1)$. So $\forall s > s_0$, $T_{e+1,s}(\emptyset) = T^{join}_{e,s}(1^m * 0)$, contradiction. So $T^{join}_e(1^m)\downarrow \forall m$. If at s_1 $T^{join}_{e,s_1}(1^m * 0)(1^m * 1)$ have settled down, then $\forall s > s_1$ $\beta_s \supseteq T^{join}_{e,s}(1^m)$. Hence $\lim_s \beta_s = \bigcup_{m \geq 0} T^{join}_e(1^m) = B$. □

(ii) Since we have B we may define T^*_m, $\forall m \leq e$, as in Lemma 3 of $\underline{m} < \underline{0}'$. Properties (1)-(6) hold.

<u>Proof</u>: We define T^*_m exactly as in Lemma 3 of $\underline{m} < \underline{0}'$ (page 35). Only property (2) needs a proof.

(2) T^*_{m+1} is either an $m+1$-splitting tree or $\exists \beta \supset B$ such that no pair of strings lying above β on T^*_{m+1} $m+1$-split β.

As in $\underline{m} < \underline{a}$ r.e. we need concern ourselves only with $T^*_{m+1} = F_{\pi(m+1)}(T^*_m)$.

Suppose we used case (i) to define T^*_{m+1}. Suppose (2) fails. Then we must have $m+1$-splittings lying on T^*_m above arbitrarily long beginnings of B.

We have, as in $\underline{m} < \underline{0}'$, an end string on T_{m+1}, $T_{m+1}(\tau) \subset \pi(m+1)$. We will show that $T_{m+1}(\tau)$ has infinitely many permanent potential extensions. A contradiction on A not recursive will then follow as in $\underline{m} < \underline{a}$ r.e.

$T_{m+1}(\tau * 0)(\tau * 1)\not\downarrow$. Hence the m+1-splittings lying on T^*_m are not permanently separated from $T_{m+1}(\tau)$ by a boundary string for T_r, $r < m+1$. If they were then we would eventually use Case III to define $T_{m+1}(\tau * 0)(\tau * 1)$.

We first show that $T_{m+1}(\tau)$ has a permanent potential extension. Since $B \supset T_{m+1}(\tau)$ there is some s_0, w and y, such that $\forall s > s_0$ $T^{join}_{w,s_0}(1^y) = T_w(1^y) \supsetneq T_{m+1,s}(\tau)$, and $T^{join}_{w,s}(1^y * 0)(1^y * 1)\downarrow$. Choose the least such $T^{join}_w(1^y)$. By hypothesis $\exists\, \sigma_1,\sigma_2$, an m+1-splitting, lying on T^*_m and $\sigma_1 \wedge \sigma_2 \supset T^{join}_w(1^y)$. Say σ_1,σ_2 lie on T^*_m $\forall s > s_1 \geqslant s_0$. Then at all stages $s > s_1$ the least such (σ_1,σ_2) is suitable as a potential extension of $T_{m+1,s}(\tau)$ and will be so used (unless $T_{m+1,s}(\tau)$ already has a potential extension). So $T_{m+1}(\tau)$ has one permanent potential extension.

If $T_{m+1}(\tau)$ has a permanent potential extension with marker $(m+1,r,t)$ then $T_{m+1}(\tau)$ has a permanent potential extension with marker $(m+1,r',t')$, $r' > r$. To show this just modify the above argument to have $lh(T^{join}_w(1^y)) > r$.

Thus $T_{m+1}(\tau)$ has infinitely many permanent potential extensions. As in $\underline{m} < \underline{a}$ r.e. we may find their associated markers recursively via T^*_m. Say they are $(m+1, r_n, s_n)$. Since we never use any, $\forall n$, $t \geq s_n \rightarrow f(t) \geq r_n$, and hence A is recursive. Contradiction.

For T^*_{m+1} defined by case (ii) the argument is similar. □

(iii) We arrive at a contradiction using (i) and (ii).

<u>Proof</u>: In (i) we showed that $T_e^{join}(1^m)\downarrow$ $\forall m$, and that $\forall s$ sufficiently large, β_s is forced above $T^{join}_{e,s}(1^m)$. We will show that this is possible only if some $e' \leq e$, some τ, $T_{e'}(\tau)$ has infinitely many permanent potential extensions. From (ii) we know that this is a contradiction.

To begin with, when β_s is forced above $T^{join}_{e,s}(1^m * 1)$ for the first time it is done so by a potential extension for some $e' \leq e$ (for one m it may be forced by g, but for all others it is not). We will show that the only potential extensions which have a permanent effect are permanent potential extensions. We explain.

Suppose we have $T_{e',s}(\tau)\downarrow$ and a potential extension of $T_{e',s}(\tau)$, (σ_1, σ_2)

forces β_s above $T^{join}_{e,s}(1^m * 1)$ for the first time.

VI-8

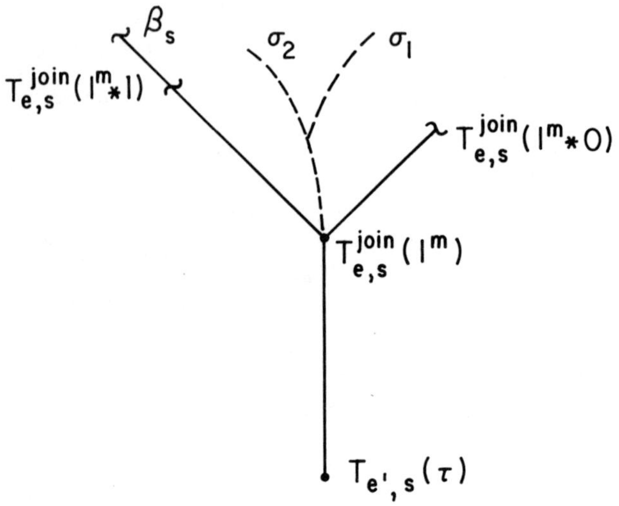

Suppose (σ_1,σ_2) is cancelled as a potential extension of $T_{e',t}(\tau)$ at some stage $t \geq s$. How can this happen? (1) $T_{e',t-1}(\tau)$ is cancelled. Or (2) we erect $T_{e',t}(\tau * 0)(\tau * 1)$ by Case II. Or (3) some ν, a boundary string for $e'' < e'$, intervenes between $T_{e',t}(\tau)$ and σ_1 or σ_2. Or (4) $\beta_{t-1} \not\supset T_{e',t}(\tau)$. Each of these cases requires us to cancel $T^{join}_{e,t-1}(1^m * 0)(1^m * 1)$. Hence the only permanent forcings are due to permanent potential extensions.

We may thus conclude that there are infinitely many permanent potential extensions associated with $T_{e'}$, some $e' \leq e$. We will show that it cannot be that $T_{e'}(\delta)$ has finitely many permanent potential extensions for infinitely many δ, but no $T_{e'}(\tau)$ has infinitely many permanent potential extensions.

Suppose this were the case. Let τ be such that $T_{e'}(\tau) \supset T^*_e(\emptyset)$ and $T_{e'}(\tau)$ has a permanent potential extension. Supppose the longest string appearing in any permanent potential extension has length r. Let δ be such that $T_{e'}(\delta)$ has a permanent extension (τ_1, τ_2) and $T_{e'}(\delta) \supset T_e^{join}(1^m)$ some m such that $lh(T_e^{join}(1^m)) > r$. We claim that (τ_1, τ_2) is suitable as a permanent potential extension of $T_{e'}(\tau)$. If the reader examines Case II (potential) he will see that the only way this could not be is if (b) failed permanently for (τ_1, τ_2) relative to $T_{e'}(\tau)$. That is, some $\sigma' \subset \tau_1$ or $\sigma' \subset \tau_2$ is a permanent boundary string for some $e'' < e'$. (see diagram next page).

VI-9

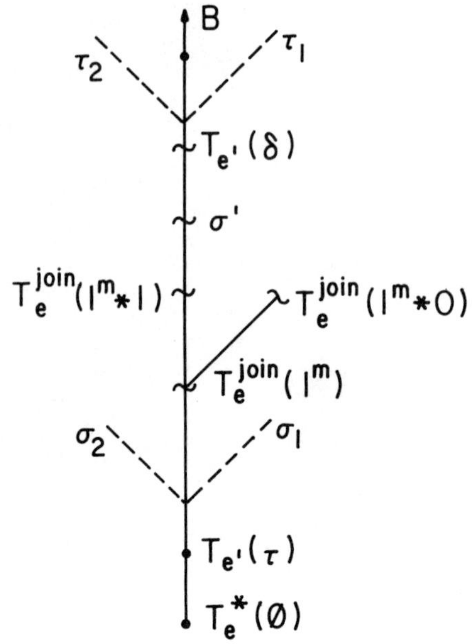

We must have $\sigma' \subsetneq T_{e'}(\delta)$ or else it would interfere with (τ_1,τ_2) relative to $T_{e'}(\delta)$. We must have $B \supset T_{e'}(\delta)$ because (τ_1,τ_2) is a permanent potential extension. But then $B \supset \sigma' = T_{e''}(\lambda)$ a boundary string for $e'' < e \supset T^*_e(\emptyset)$. This cannot be: T^*_e is compatible with $T_{e''}$ so σ' must lie on T^*_e; boundary strings are end strings on T^*_e; but B lies on T^*_e, contradiction.

So (τ_1,τ_2) is suitable as a permanent potential extension of $T_{e'}(\tau)$ and will be so used. Contradiction on the longest string in any permanent potential extension of $T_{e'}(\tau)$ having length r.

Hence some $e' \leqslant e$, some τ, $T_{e'}(\tau)$ has infinitely many permanent potential extensions, our final contradiction. □

Corollary: $\lim_s t(e,s)\downarrow$.

Proof: $T_{e+1}(\emptyset) = T_e^{join}(1^m * i)$ some $m \geq 0$, $i \in \{0,1\}$. Then $\lim_s t(e,s) = lh(T_e^{join}(1^m * 1))$. □

Lemma C) $\underline{a} \cup \underline{b} = \underline{0}'$.

Proof: Since $\underline{b} \leq \underline{0}'$, $\underline{a} \cup \underline{b} \leq \underline{0}'$. We will show that $\underline{0}' < \underline{a} \cup \underline{b}$ by giving a computation procedure for K recursive in $A \oplus B$.

Let $r(e) = \mu s(s > r(e-1)$ and $\forall x \leq t(e,s)$, $A^s(x) = A(x) \wedge \beta_s(x) = B(x))$. Since $\lim_s t(e,s)\downarrow$, $r(e)\downarrow \ \forall e$. $r(e) \leq_T A \oplus B$.

We claim that $e \in K \leftrightarrow e \in K^{r(e)}$. Then $K \leq_T A \oplus B$.

Suppose our claim is wrong. Then some e, $e \in K$ but $e \notin K^{r(e)}$. Choose the minimal such e. Then $g(w) = e$ some $w > r(e)$.

At stage $r(e)$, $t(e,r(e))\downarrow$; say $t(e,r(e)) = lh(T_{e,r(e)}^{join}(1^m * 1))$. We must have $\beta_{r(e)} \supseteq T_{e,r(e)}^{join}(1^m * 0)$. If we did not then $t(e,r(e))$ is not defined with respect to this branching, for we cannot have made a switch for g at e.

$\forall x \leq t(e,r(e))$, $A^{r(e)}(x) = A(x)$. Hence no permitting is allowed below $T_{e,r(e)}^{join}(1^m * 1)$ at any stage $s > r(e)$.

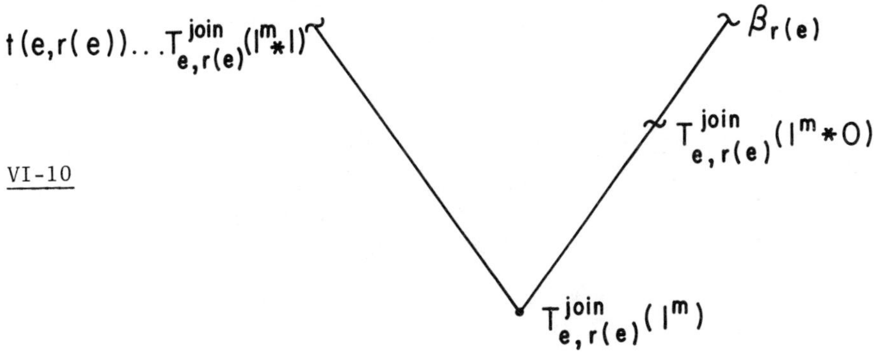

VI-10

We have two possibilities. We show that both lead to contradictions.

(1) Some $s > r(e)$, s minimal, $T^{join}_{e,s-1}(1^m * 0)(1^m * 1)$ is cancelled. This cannot be due to an application of permitting. It must be due to β_s now being forced above $T^{join}_{e',s}(1^n * 1)$, some $e' < e$, some n, and $\beta_{s-1} \supset T^{join}_{e',s-1}(1^n * 0)$. Then $\beta_{r(e)} \supseteq T^{join}_{e,r(e)}(1^m * 0) \supset T^{join}_{e',s-1}(1^n * 0)$.

VI-11

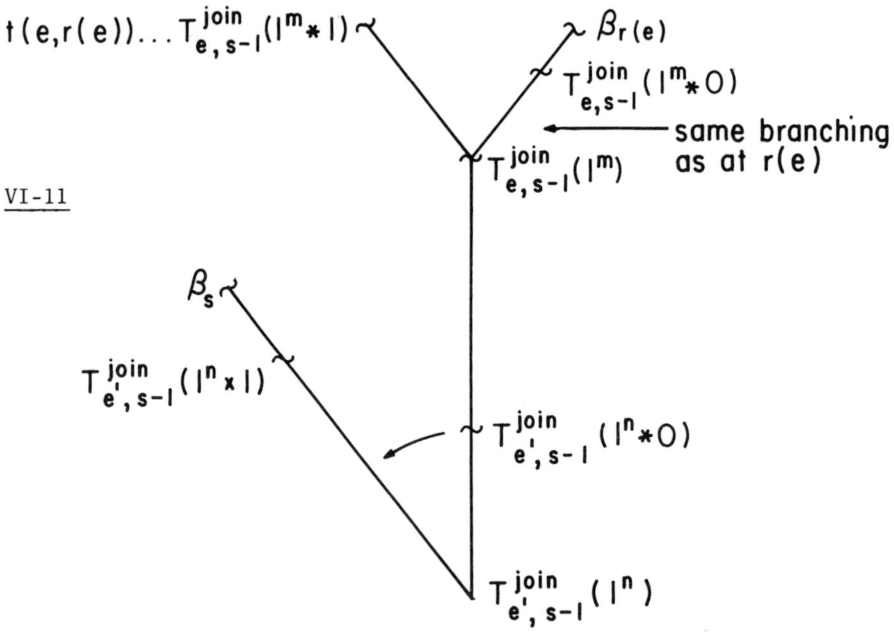

Is it possible for $\beta_t \supseteq T^{join}_{e',s-1}(1^n * 0)$, $t > s$? Since there is no permitting allowed below $t(e,r(e))$ there is none allowed below $lh(T^{join}_{e',s-1}(1^n * 0))$. Hence the only way for $\beta_t[lh(T^{join}_{e',s-1}(1^n * 0))] \neq$

$\beta_{t-1}[lh(T^{join}_{e',s-1}(1^n * 0)]$ is by a repetition of the above situation for some $e'' < e'$, some n''. That clearly cannot place $\beta_t \supseteq T^{join}_{e',s-1}(1^n * 0)$. So $\beta_{r(e)} \supseteq T^{join}_{e',s-1}(1^n * 0)$, $B \not\supseteq T^{join}_{e',s-1}(1^n * 0)$. $lh(T^{join}_{e',s-1}(1^n * 0)) < t(e,r(e))$. Contradiction on the definition of $r(e)$.

(2) $\forall s > r(e)$, $T^{join}_{e,s}(1^m * 0)(1^m * 1) = T^{join}_{e,r(e)}(1^m * 0)(1^m * 1)$. Recall that $g(w) = e$ some $w > r(e)$. If β_w is not already forced above $T^{join}_{e,w}(1^m * 1) = T^{join}_{e,r(e)}(1^m * 1)$ it is now forced by g. β_s is then forced above $T^{join}_{e,s}(1^m * 1) = T^{join}_{e,r(e)}(1^m * 1)$ $\forall s > w$, since this branching is never cancelled. Hence $B \supset T^{join}_{e,r(e)}(1^m * 1)$. $\beta_{r(e)} \supseteq T^{join}_{e,r(e)}(1^m * 0)$. Contradiction on the definition of $r(e)$.

Hence no such e exists, and $\forall e$, $e \in K \leftrightarrow e \in K^{r(e)}$. □

This completes the proof of the theorem.

Corollary: $\underline{a}$ r.e. $\to$ $\underline{a}$ has a complement in the upper semi-lattice of degrees $\leq \underline{0}'$.

Proof: For $\underline{a} = \underline{0}$ or $\underline{a} = \underline{0}'$ it is immediate. For $\underline{0} < \underline{a} < \underline{0}'$ we have a minimal $\underline{m} < \underline{0}'$ such that $\underline{m} \cup \underline{a} = \underline{0}'$. $\underline{m} \cap \underline{a} = \underline{0}$ because $\underline{m}$ is minimal. □

Further Topics and Remarks

It is possible to generalize the idea of permitting to arbitrary $\underline{a}$, $\underline{0} < \underline{a} \leqslant \underline{0}'$. To do this we look at $A \in \underline{a}$, $\lim_s A_s(x) = A(x)$, $\{A_s\}_{s \geqslant 0}$ uniformly recursive, and set

$$C_A(x) = \mu s(s > C_A(x-1),\ A_s[x] = A[x])$$

If A is not recursive then C_A is not dominated by any recursive function, (first noted in (5)).

Proof: If ϕ recursive dominates C_A then to compute $A(x)$ recursively look for the least n such that $x < n$ and $A_s(x) = A_{\phi(n)}(x)$ for each s, $\phi(n) \leqslant s \leqslant \phi(\phi(n))$. Then $A(x) = A_{\phi(n)}(x)$, for $C_A(\phi(n))$ lies between $\phi(n)$ and $\phi(\phi(n))$.

If we define

$$C_A(s,x) = \mu t(s \geqslant t > C_A(s,x-1),\ A_t[x] = A_s[x]),$$

then $\lim_s C_A(s,x) = C_A(x)$. We can then use $C_A(s,x)$ to tell us when to permit something. That is, instead of asking for $f(u) < r$ to permit the use of something with marker (e,r,t) we will ask for $C_A(u,r) > t$; that is, $A_t[r]$ has not yet read correctly. Of course $C_A(u,x)$ may be reading different from $C_A(x)$ so we will change our minds a lot as to whether to permit something or not.

R.W. Robinson (6) has used this generalized permitting to show that $\underline{0} < \underline{a} \leqslant \underline{0}' \rightarrow \exists \underline{b} < \underline{0}'$ such that $\underline{a} \cup \underline{b} = \underline{0}'$. From this we have that $\underline{m} < \underline{0}'$, $\underline{m}$ minimal $\rightarrow \underline{m}$ has a complement in the degrees $\leqslant \underline{0}'$.

He (and we independently) then showed how to modify his proof to get $\underline{0} < \underline{a} < \underline{0}'$ and $\underline{a}'' = \underline{0}'' \rightarrow \exists \underline{b} < \underline{0}'$ such that $\underline{b} \cup \underline{a} = \underline{0}'$

and $\underline{b} \cap \underline{a} = \underline{0}$. That is, $\underline{b}$ has a complement in the degrees $\leq \underline{0}'$. This proof also appears in (6). It is based on the fact that $\underline{a} < \underline{0}'$ and $\underline{a}'' = \underline{0}'' \to \{\underline{d} : \underline{d} \leq \underline{a}\}$ is recursive in $0'$. This latter means $\exists$ a recursive functional of two variables, Φ, such that $\Phi_e(0',n)(x)\downarrow$ $\forall n,x$ and $\underline{d} \leq \underline{a} \to \exists n$, $\Phi_e(0',n)(x) \in \underline{d}$. This fact allows us to diagonalize $\{\underline{c} : \underline{c} \leq \underline{b}\}$ against $\{\underline{d} : \underline{d} \leq \underline{a}\}$ in the proof of $\underline{b} \cup \underline{a} = \underline{0}'$ to get $\underline{b} \cap \underline{a} = \underline{0}$.

In (2) Cooper has shown that $\underline{h}$ high, that is $\underline{h} < \underline{0}'$ and $\underline{h}' = \underline{0}''$, $\to$ there is a minimal degree $\underline{m} < \underline{h}$. To get this he uses the fact that $\underline{h}$ high $\to$ $\exists A \in \underline{h}$ such that A has an approximation for which $C_A(x)$ dominates every recursive function. We believe that it is possible to use this "high-permitting" to prove:

> $\underline{h}$ high $\to$ there is a minimal degree $\underline{m} < \underline{0}'$
> such that $\underline{m} \cup \underline{h} = \underline{0}'$.

Cooper's proof would need to be modified along the same lines as $\underline{m} < \underline{a}$ r.e. was to get $\underline{m} \cup \underline{a}$ r.e. $= \underline{0}'$.

We have outlined all the classes of degrees $< \underline{0}'$ for which we know complements exist: $\{\underline{a} : \underline{a}'' = \underline{0}''\}$, $\{\underline{a} : \underline{a}$ r.e.$\}$, $\{\underline{m} : \underline{m}$ minimal$\}$ and possibly $\{\underline{h} : \underline{h}$ high$\}$. Here we are face to face with the nature of the study of the degrees $\leq \underline{0}'$ so far. These classes, plus $\{\underline{a} : \underline{a}' = \underline{0}'\} \subseteq \{\underline{a} : a'' = \underline{0}''\}$ are the only classes for which we have characterizations which are useful in general constructions. Perhaps this list will have to be expanded before we can answer the question:

> Is there a $\underline{b} < \underline{0}'$ such that $\underline{b}$ does not have a complement in the degrees $\leq \underline{0}'$?

Another question is suggested by the result of this chapter:

> Is there a $\underline{b} < \underline{0}'$ such that $\underline{b}$ has a complement in the degrees $\leqslant \underline{0}'$, but none which is a minimal degree?

Cooper (1) has shown a very interesting fact, namely that there are two minimal degrees $\underline{m}_1, \underline{m}_2 < \underline{0}'$ such that $\underline{m}_1 \cup \underline{m}_2 = \underline{0}'$. Is it possible that every minimal degree $\leqslant \underline{0}'$ has a complement in the degrees $\leqslant \underline{0}'$ which is also a minimal degree?

In (1) Cooper also gives a good proof of the simple existence of two degrees complementary below $\underline{0}'$. We can modify that proof to show that there are two high degrees complementary below $\underline{0}'$; and that there are two degrees $\underline{a}_1, \underline{a}_2$ such that $\underline{a}_1' = \underline{a}_2' = \underline{0}'$ which are complementary below $\underline{0}'$. We believe it is possible to show that given $\underline{c}_1, \underline{c}_2$ r.e. in $\underline{0}'$, there are $\underline{a}_1, \underline{a}_2$ complementary below $\underline{0}'$ such that $\underline{a}_1' = \underline{c}_1$ and $\underline{a}_2' = \underline{c}_2$. Thus finding complements for degrees $\leqslant \underline{0}'$ does not appear to be limited in any way by the jump operator.

Along these lines the reader may note that we can apply the methods of Chapter III, $\underline{m}' = \underline{0}'$, to the construction of this chapter to obtain that the minimal degree that we construct satisfies $\underline{m}' = \underline{0}'$.

BIBLIOGRAPHY

[1] Cooper, S.B. Degrees of unsolvability complementary between recursively enumerable degrees. Annals of Mathematical Logic, vol. 4, no. 2, (1972).

[2] Cooper, S.B. Minimal degrees and the jump operator. Journal of Symbolic Logic, vol. 38, no. 2 (1973).

[3] Friedberg, R.M. A criterion for completeness of degrees of unsolvability. Journal of Symbolic Logic, vol. 22 (1957).

[4] Lachlan, A.H. Lower bounds for pairs of recursively enumerable degrees. Proceedings of the London Mathematical Society, (3) 16 (1966).

[5] Miller, W. and Martin, D.A. The degrees of hyperimmune sets. Zeitschrift für Mathematische Logik und Grundlagen der Mathematik, 14 (1968).

[6] Robinson, R.W. Degrees joining to $\underline{0}'$. (to appear)

[7] Rogers, H. Theory of Recursive Functions and Effective Computability, McGraw-Hill, San Francisco, 1967.

[8] Sacks, G.E. On the degrees less than $\underline{0}'$. Annals of Mathematics, 77 (1963).

[9] Sasso, L. A minimal degree not realizing least possible jump, Journal of Symbolic Logic (to appear).

[10] Shoenfield, J.R. On degrees of unsolvability. Annals of Mathematics, 69 (1959).

[11] Shoenfield, J.R. Degrees of Unsolvability, North-Holland, London, 1971.

[12] Spector, C. On degrees of recursive unsolvability, Annals of Mathematics, vol. 64 (1956).

[13] Yates, C.E.M. Initial segments of the degrees of unsolvability, Part II: Minimal degrees. Journal of Symbolic Logic, vol. 35, no. 2 (1970).

INDEX

Notations:

Definitions:

Notes added in proof :

1. It should be noted that there was another full approximation style construction extant prior to Cooper's : namely due to Yates. It was needed, of course, for Yates to prove that given $\underline{0} < \underline{a}$ r.e.$\exists$ $\underline{m}$ minimal $\underline{m} < \underline{a}$. That we refer to this construction only in passing is not intended to denigrate its importance. Rather it reflects our preoccupation with the construction with which we have chosen to work.

2. p vi, l b5 is intended of course to refer only to published works. We realize that diagrams have been used in class work by several people, and indeed our diagrams are derived from this tradition.

3. p 79. Permitting has a long history and our intent has been only to indicate its content here. For a detailed account of permitting the reader should consult Yates [13].

4. D. Posner has recently pointed out that in constructing any minimal degree $\underline{m}$ it is not necessary to diagonalize to show that $\underline{m} \neq \underline{0}$.
viz :

Suppose $\forall n$ B lies on a partial recursive tree T_n such that T_n is an n-splitting tree or has no n-splittings. Then B is not recursive.

Proof : Suppose not; then define

$$\Phi(A)(x) = \begin{cases} A(x) & \text{if } \exists y\ A(y) \neq B(y) \\ \uparrow & \text{otherwise} \end{cases}$$

Then Φ is partial recursive and hence

$\Phi = \Phi_n$ some n.

T_n cannot be an n-splitting tree, since B lies on T_n and $\forall x,\ \Phi_n(B)(x) \uparrow$. But then T_n must have no n-splittings. Yet for any σ_1, σ_2 lying on T_n such that $\sigma_1 | \sigma_2$ and $\sigma_1, \sigma_2 \subseteq B$ we have some x, $\Phi_n(\sigma_1)(x) \neq \Phi_n(\sigma_2)(x)$, an n-splitting. Hence B is not recursive. □

This fact simplifies considerably the construction of Chapter V, making the trees $T^{rec}_{e,s}$ unnecessary. Nonetheless, we urge the reader to understand the technique of $T^{rec}_{e,s}$ as it is the simplest prototype of the more difficult $T^{join}_{e,s}$ used in Chapter VI, which is certainly non-trivial.

5. D. Posner has recently answered a question on p 129; namely, he has shown that

$$\underline{h} \text{ high} \rightarrow \quad \text{there is a minimal degree } \underline{m} < \underline{0}' \text{ such that } \underline{m} \cup \underline{h} = \underline{0}'.$$

This as well as further important work on high degrees appears in his doctoral dissertation at the University of California, Berkeley.

6. We have recently shown that all the facts concerning joins and complements in the degrees $\leq \underline{0}'$, which we discuss at the end of Chapter VI, hold as well in the degrees $\leq \underline{h}$, where $\underline{h}$ is any high degree. This will appear in our forthcoming paper "Joins and Complements Below High Degrees".

235 Montgomery St., Suite 450
San Francisco
California 94104
U.S.A.